THE TABOO OF SUBJECTIVITY

○ ○ ○

Toward a New Science of Consciousness

B. Alan Wallace

OXFORD

UNIVERSITY PRESS

OXFORD
UNIVERSITY PRESS

Oxford   New York
Auckland   Bangkok   Buenos Aires   Cape Town   Chennai
Dar es Salaam   Delhi   Hong Kong   Istanbul   Karachi   Kolkata
Kuala Lumpur   Madrid   Melbourne   Mexico City   Mumbai   Nairobi
São Paulo   Shanghai   Taipei   Tokyo   Toronto

First published in 2000 by Oxford University Press, Inc.
198 Madison Avenue, New York, New York 10016
First issued as an Oxford University Press paperback, 2004

www.oup.com

Oxford is a registered trademark of Oxford University Press.

Library of Congress Cataloging-in-Publication Data
Wallace, B. Alan.
The taboo of subjectivity : toward a new science of consciousness
/ B. Alan Wallace.
p.   cm.
Includes bibliographical references and index.
ISBN-13 978-0-19-517310-9

1. Religion and science.   2. Materialism.   3. Consciousness.
I. Title.
BL240.2.W27   2000
291.1'75—dc21   99-44840

Printed in the United States of America

THE TABOO OF SUBJECTIVITY

To my parents and teachers,
who instilled in me the faith to question.

ACKNOWLEDGMENTS

This work has benefited greatly from the critical comments I have gratefully received from Lee Yearley, Hester Gelber, and Van Harvey of the Religious Studies Department at Stanford University. I am also indebted to Robert Livingston, Professor Emeritus of Neuroscience at the University of California at San Diego; to Mark Sullivan, of the Department of Psychiatry and Behavioral Sciences at the University of Washington; Anne Harrington, of the Department for the History of Science at Harvard University; Eugene Taylor, of the Cambridge Institute of Psychology and Religion; Francisco Varela, Director of Research at the Centre Nationale de la Recherche Scientifique (CNRS), Paris; Piet Hut, of the Institute for Advanced Study at Princeton University; Arthur Zajonc, of the Department of Physics at Amherst College; Amit Goswami, of the Department of Physics at the University of Oregon; David Finkelstein, of the School of Physics at Georgia Institute of Technology; Anton Zeilinger, of the Institut für Experimentalphysik at the University of Vienna; Greg Simpson, of the Department of Neurology at Albert Einstein College of Medicine; P. Luigi Luisi, of the Swiss Federal Institute of Technology; Thomas J. McFarlane, of the California Insitute of Integral Studies; Lynn Quirolo; and Calvin Smith. I especially wish to express my deep gratitude to David Galin, of the Center for Social and Behavioral Sciences at the University of California at San Francisco, for his insightful critique of this manuscript. In addition, I wish to extend my thanks to the Fetzer Institute, which provided me with a grant to enable me to complete this work. I am also very grateful to Cynthia Read, Executive Editor for Religion at Oxford University Press, for her unflagging support of this work, and to Robert Milks, also of Oxford University Press, and Martha Ramsey for helping to polish it into its final form.

Finally, I can never repay the kindness of my wife, Vesna A. Wallace, of the Department of Religious Studies at the University of California, Santa Barbara; my parents, Barbara and David H. Wallace; and my mentors Geshe Rabten, Gyatrul Rinpoche, and H. H. the Dalai Lama of Tibet for arousing my mind to important questions and helping me in my pursuit of deepening understanding.

I have benefited greatly from my collaboration with these scholars, though, of necessity, I have not been able to implement all their suggestions or insights. Thus, whatever errors remain in the text are my sole responsibility, and I hope that my further studies and collaboration with scholars, scientists, and contemplatives more learned and experienced than me will bring them to my attention.

Santa Barbara, California   B. Alan Wallace
January 2000

CONTENTS

Introduction: The No Man's Land of Consciousness   3

PART I *The Ideology of Scientific Materialism*

1 Four Dimensions of the Scientific Tradition   17

2 Theological Impulses in the Scientific Revolution   41

PART II *Toward a Noetic Revolution*

3 An Empirical Alternative   59

4 Observing the Mind   75

5 Exploring the Mind   97

PART III *The Resistance*

6 The Mind in Scientific Materialism   123

7 Confusing Scientific Materialism with Science   145

8 Scientific Materialism: The Ideology of Modernity   159

Conclusion: No Boundaries   177

Notes   189

Bibliography   197

Index   209

THE TABOO OF SUBJECTIVITY

INTRODUCTION
The No Man's Land of Consciousness

When we consider what religion is for mankind, and what science is, it is no exaggeration to say that the future course of history depends upon the decision of this generation as to the relations between them.

Alfred North Whitehead

Among all the points of contact between science and religion, there is none more crucial and none more clouded in mystery and confusion than the views concerning the nature of consciousness. While many philosophers acknowledge that little or nothing is known about consciousness, many people today make strong, diverse claims concerning the human soul and consciousness based upon religious and scientific authority. Religious believers interpret consciousness in accordance with their respective creeds, the authority of which is not accepted by many scientists; and scientists base their views of consciousness on the metaphysical principles underlying scientific inquiry, the validity of which is questioned by many religious believers.

Despite centuries of modern philosophical and scientific research into the nature of the mind, at present there is no technology that can detect the presence or absence of any kind of consciousness, for scientists do not even know what exactly is to be measured. Strictly speaking, *at present there is no scientific evidence even for the existence of consciousness!* All the direct evidence we have consists of nonscientific, first-person accounts of being conscious. The root of the problem is more than a temporary inadequacy

3

of the technology. It is rather that modern science does not even have a theoretical framework within which to conduct experimental research.[1] While science has enthralled first Euro-American society and now most of the world with its progress in illuminating the nature of the external, physical world, I shall argue that it has eclipsed earlier knowledge of the nature of the inner reality of consciousness. In this regard, we in the modern West are unknowingly living in a dark age. A central aim of this book is to unveil the ideological constraints that have long been impeding scientific research in the study of consciousness and other subjective mental states.

As an illustration of the standoff between science and Christianity, the dominant religion in the West, consider first the question of the origins of human consciousness. Many Christians firmly believe that a human fetus is endowed with a human soul or consciousness from the moment of conception. Many other people, relying on scientific understanding of human embryology, are equally convinced that at least during the first and perhaps the second trimester the fetus is not conscious. A limitation of both these views is that neither is based on compelling empirical evidence. Christians base their positions solely on the authority of their own tradition, but they are unable to demonstrate the validity of their views to anyone who does not share their faith. Augustine (354–430), a theologian whose thinking has had an enormous impact on both Roman Catholic and Protestant theology, declared that the problem of the origin of the human soul remained a mystery to him due to its "depth and obscurity." This subject, he claimed, had not been studied sufficiently by Christians to be able to decide the issue, or if it had, such writings had not come into his hands. While he suspected that individual souls are created under the influence of individual conditions present at the time of conception, he acknowledged that, as far as he knew, the truth of this hypothesis had not been demonstrated.[2] Instead of seeking compelling empirical evidence concerning the origins of consciousness, the Christian tradition has drawn its conclusions around this issue on purely doctrinal grounds. But, according to Augustine, it is an error to mistake mere conjecture for knowledge.

It is as difficult to determine the basis of the scientific view. Modern science does not know any better than Augustine how or why consciousness originates, nor does it have any way of directly detecting the presence or absence of consciousness in a human fetus or even a human adult. In the absence of any compelling evidence, advocates of this view have simply formed an opinion and asserted that as their orthodox view. But there is little to distinguish religious ignorance from scientific ignorance.

We encounter a similar dilemma in terms of Christian and scientific views of the nature of consciousness during the course of human life. Christians commonly claim that each of us is endowed with free will, which makes us morally responsible for our actions. Acting as free agents, we make decisions, sometimes after a good deal of internal struggle, and we must take responsibility for the results of our choices and deeds. Many advocates of science, on the other hand, claim that all mental behavior is produced strictly by the brain's response to physical stimuli in accordance with the laws of nature. In this view, our subjective sense of making choices, intentionally pursuing our desires, and acting on the basis of our beliefs is illusory in the sense that our actions are in reality simply products of our brains in interaction with the environment.

Regarding this issue, which is central to the definition of the very nature of human existence, Christians again call on the authority of their tradition. Advocates of science, on the other hand, bolster their position by pointing to a growing body of neuroscientific knowledge of correlates between specific brain functions and specific mental processes. Neuroscientists have discovered that when certain functions of the brain are altered or inhibited, specific mental functions change or cease altogether. Such empirical evidence suggests that those mental functions are *conditioned* by their respective brain functions, but it does not rule out in principle the possibility of other, possibly nonphysical, factors influencing the mind. Thus, the scientific evidence alone does not compel us to believe that the brain is *solely* responsible for the creation of *all* conscious states.

As for the nature of death, most religions, including Christianity, assert the continuity of individual consciousness following this life, and the authority of sacred scriptures is invoked to substantiate this claim. Mainstream neuroscience, in contrast, insists that individual consciousness vanishes with the death of the body. However, given its ignorance of the origins and nature of consciousness and its inability to detect the presence or absence of consciousness in any organism, living or dead, neuroscience does not seem to be in a position to back up that conviction with empirical scientific evidence. It is remarkable that despite the many diverse branches of science that explore every aspect of the known universe, we still have no science of consciousness, only philosophical and religious beliefs. So I am left with the question: Can science provide an adequate view of the entire natural world that includes only objective phenomena, while excluding the subjective phenomenon of consciousness altogether?

In short, however deeply we may hold to our present religious or scientific convictions concerning such issues as free will and the possibility of an afterlife, there are large gaps in our knowledge about the one phenomenon that holds the key to these questions. That phenomenon is our own consciousness, about which the *International Dictionary of Psychology* asserts: "[c]onsciousness is a fascinating but elusive phenomenon: it is impossible to specify what it is, what it does, or why it evolved. Nothing worth reading has been written about it."[3] This statement exemplifies much modern Western thinking on this topic; while the author of this statement acknowledges, at least implicitly, that he does not understand consciousness, he simultaneously declares that no one else does either and that it is impossible to understand. On the contrary, I would argue that consciousness is not impossible to specify, and much has been written about it that is eminently worth reading. By the term "consciousness" I mean simply the sheer events of sensory and mental awareness by which we perceive colors and shapes, sounds, smells, tastes, tactile sensations, and mental events such as feelings, thoughts, and mental imagery. Thus, I am using the word "consciousness" to refer to the *phenomenon* of being conscious, not to the neural events that make this first-person experience possible.

While consciousness lies in the no man's land between religion and science, claimed by both yet understood by neither, it may also hold a key to the apparent conflict between these two great human institutions. This is a second theme that weaves itself throughout this work. To place our individual perspectives in context, it may be helpful to note that according to recent polls, between 70 percent and 90 percent of all Americans believe in a personal God, 80 percent believe in angels, and 40 percent believe that God has guided the evolution of life; while only 9 percent believe that God had no part in human development over millions of years from less advanced forms of life. Moreover, 40 percent of the American scientists polled acknowledge their belief in a personal God to whom they can pray, which is roughly the same percentage as in a poll taken a century ago.[4] On the other hand, according to other recent polls, 10 percent of the German population still believes in a stable earth, and a third of all adults in the United States believe everything in the Bible to be literally true.[5] This range of statistics indicates that a significant minority of people in the West simply dismiss scientific knowledge and another significant minority simply dismiss religious beliefs, but a majority of people in the modern West are caught up in the conflict between science and religion.

Those of us who find ourselves in this middle ground generally try to reconcile the domains of religion and science by separating them in various ways, and such attempts have been going on over the past four centuries. Since the Scientific Revolution of the sixteenth and seventeenth centuries, most scientists, beginning with Copernicus, Galileo, Descartes, and Newton, have sought to accommodate their scientific theories to the orthodox theologies of their times. Even in the nineteenth century, most British "men of science" still thought that there was no essential conflict between their science and those parts of the Christian faith that liberal Christians still regarded as essential.[6]

This way of thinking is in evidence among contemporary intellectuals, even those in the cognitive sciences, which often strike at the heart of religious belief. To take but one example, some professionals in this field have suggested that Christians can adopt modern scientific, physicalist theories of the mind and still hold to an orthodox Christian belief in eternal life. Their proposal is that Christians may accept the scriptural promise of life after death, even if people do not survive the total destruction of the body; for they can still look forward to everlasting life in the sense that ordinary death does not entail the final dissolution of the body. Presumably such thinkers are referring to the resurrection of the body when it is transformed at the time of Christ's return. This view, of course, does not allow for the existence of a soul that continues to exist independently of the body.[7]

Most Christians are understandably disinclined to accept this physicalist view of immortality, which they regard as incompatible with the Bible.

If we are to hold religious beliefs and to accept scientific progress, how are we to draw the line between the domains of these two views? One possibility is to look solely to religion to clarify the fundamental ends and value standards of human endeavors, and to look solely to science for genuine knowledge of the nature of reality.[8] This solution appears to me inadequate, for the ideals and values of religion are based on religious statements about the nature of reality. Christian values, for example, are based on assertions of the truth of God's existence, the immortality of the soul, the power of prayer, and so on. Indeed, if one accepts the truth of the Christian worldview, Christian values and ideals must be accepted as a matter of course; but if that worldview is rejected, the Christian rationale for those values and ideals is undermined.

On the other hand, a thoroughly materialistic view of the universe based on science suggests quite a different set of values and ideals, with profound implications for dealing with the personal, societal, and environmental problems that beset us today. The attempt to embrace religious ideals while adhering to a thoroughly materialistic worldview is severely hampered from the outset. What we really believe to be true will invariably influence what we believe to be of value; conversely, all of us, including scientists, seek to understand those aspects of reality that we value. Thus, the scientific worldview has been generated by the kinds of values and ideals held by scientists. The mutual interdependence of values and beliefs is inescapable.

According to some thinkers, a more feasible way of demarcating science and religion is to grant science authority in terms of knowledge of the natural world and to appoint it the task of providing humanity with the technological means of mastering the forces of nature to ensure our physical survival and well-being. The proper arena of religion, they say, is the sacred world, with all the ideals and moral directives for human behavior that issue forth from that domain.[9] This model is feasible if we believe the sacred world exists independently of, and has no influence in, the world of nature and human life. But the great majority of religious believers today believe that the object, or objects, of their religious devotion is very much present and active in nature and in the lives of human beings. Thus, according to those believers, the absolute demarcation between the sacred and the profane is untenable.

Another approach to this problem is to distinguish science from religion in terms not of their domains of authority but their methodologies. Following this line of thought, science may be identified by its methodology of *depersonalizing* phenomena. That is, science attempts to account for a given phenomenon independently of the particular subject who observes it. Religion, on the contrary, some argue, is based on experiences taken in their subjective and individual elements.[10] As a result of the disparity between these two methodologies, by the nineteenth century the relation between science and religion had become one of radical dualism. Each of them was regarded as absolute and as utterly distinct as, according to the reigning psychology of that day, the two faculties of the soul—intellect and feeling—to which they respectively corresponded. But our intellect and feelings do not function autonomously; our thoughts are frequently charged with emotion, and our feelings arise in response to what we think to be true. To reify and alienate these facets of our inner life is to fragment each of us from within. We are persons whose bodies can be objectively studied according to the impersonal laws of physics but whose minds are subjectively experienced in ways science has not yet been able to fathom. In short, by radically separating science from religion, we are not merely segregating two human institutions; we are fragmenting ourselves as individuals and as a society in ways that lead to deep, unresolved conflicts in terms of our view of the world, our values, and our way of life.[11]

To bring my own background into this discussion, this was the situation in which I found myself in the late 1960s as an undergraduate student of biology at the University of California. Through my childhood and youth, I had been raised in a devout, educated Christian family and was encouraged by my theologian father to pursue my interest in a career in science. The mainstream Protestant Christianity to which I had been exposed presented itself as an integrated and comprehensive worldview, value system, and way of life in accordance with the Bible; and this was advocated as the one true religion, the sole means to personal salvation. But I was clearly aware that other religions around the world and throughout history had long been making similar dogmatic claims that *they* were the one true faith. Whether one accepted Christianity, Judaism, Islam, or Buddhism as providing a uniquely true picture of reality and the one way to salvation seemed to me primarily an accidental matter of the time and place of one's birth. This was hardly a compelling reason for me to believe any of these exclusivist claims.

On the other hand, the scientific knowledge to which I had been exposed presented itself as an integrated and comprehensive worldview, which had its own implicit set of values and ideals for human life. Moreover, much of the secular education I received asserted that scientific progress had from the beginning been only impeded by religion and that religious beliefs had on the whole been discredited by scientific knowledge. Thus, any truly educated person, I was told, must see that scientific authority had displaced religious authority and that science alone provided a uniquely true picture of reality and alone should be relied on to solve the broad range of problems confronting humanity.

As a young man aspiring to a career in environmental studies, I found this exclusivist position of scientists just as unsatisfactory as similar claims of religious believers. Global pollution, rapid depletion of natural resources, the population explosion, the extinction of more and more species of plant and animal life, and the proliferation of nuclear, conventional, chemical, and biological weapons were just a few of the enormous problems for which purely scientific and technological solutions were obviously inadequate. These problems were created not simply by lack of scientific knowledge—indeed many of them would not have occurred without scientific knowledge—but by such human vices as greed, aggression, and shortsightedness. Even if scientists found effective ways to solve these problems, what was to persuade society to make the necessary sacrifices to implement them? How could the industrialized societies be persuaded to stop consuming the lion's share of the world's resources? And how could the developing nations of the world be persuaded not to desecrate the natural environment in emulation of the more affluent, technologically advanced nations? External environmental problems that imperil our very existence as a species appear to be the result of internal human problems, and if those internal issues are not addressed, no external solution can be effectively implemented.

While scientific knowledge alone seemed to me as a young man to be insufficient for dealing with global problems, I was also struck by its inadequacy for addressing my own personal aspirations and conflicts. It was obvious to me that one could be well educated, affluent, and living in good health in a comfortable environment and still be tense, anxious, and dissatisfied. Many people who find themselves in that situation simply fall into depression, for which modern medicine prescribes powerful drugs to alleviate their symptoms. But the scientific discipline of psychiatry has no formula or prescription for finding an inner sense of meaning, contentment, or fulfillment. I do not find it at all surprising that many people in the modern West turn in desperation to illegal drugs in their pursuit of happiness. Science has encouraged us to look outward for solutions to all our problems, social and personal, and illegal drugs are seen as just one more option.

To continue my own personal narrative, which underlies and motivates this book, at the age of twenty-one I became so disenchanted with all the options presented to me by both my religious and scientific education that I turned my back on my native society to seek out the wisdom of a civilization radically dissimilar to, and disengaged from, the modern West. This quest brought me to Dharamsala, India, the home of the Dalai Lama and a nucleus of Tibetan civilization in exile. While I had voluntarily exiled myself from my homeland in disillusionment with my native culture, the Tibetan refugees with whom I came to live had fled their beloved homeland in order to preserve their native culture.

In 1949, the Chinese communists had invaded Tibet, and in the ensuing decades, especially during the Cultural Revolution, they were responsible for the deaths of up to a million Tibetans and the genocidal destruction of Tibetan civilization. The heart of this culture is Tibetan Buddhism, and to this day it has been the special target of Chinese aggression. This indicates that at present the main ideological thrust of the Chinese mission in Tibet is not the imposition of a socialist economic system—which the Dalai Lama and many other Tibetans are happy to embrace—but a belief system that is profoundly at odds with the worldview of Tibetan Buddhism. That ideology is still being forcefully propagated in Tibet, long after the socialist ideals of Mao have been displaced by the capitalist ideals of unbridled industrial development and consumerism.

What then is this ideology? According to the current Chinese propaganda in Tibet, it is the doctrine that science presents the one true view of reality and the solutions to humanity's problems are all to be found in technology. Religion, this doctrine declares, is superstition, and it must be rooted out by whatever means necessary, including forcible indoctrination and violence. While it would be wonderful for the Tibetans to learn about science and technology—and the Dalai Lama himself is keenly interested in such knowledge and is strongly backing science education for the Tibetans in exile—the Chinese are intent on promoting a materialistic ideology more than on promoting science itself. Thus, I have found that when backed by political and military power without restraint by the ideals of democracy, the ideology of science can be just as intolerant and vicious in its suppression of competing worldviews as any traditional religion. Moreover, while Tibetans had for centuries maintained a sustainable economy and population in balance with their natural environment, since the Chinese invasion, Tibet has been largely denuded of its forests, its wildlife has been ravaged, and its cities have been polluted; its northern plateau is now used as a dumping site for nuclear waste. Instead of having the opportunity of a liberal education in modern science, the Tibetans have been hammered by the iron fist of a science-based ideology that has suppressed freedom of thought and led to the desecration of their homeland.

What immediately struck me about the Tibetan refugees in Dharamsala, despite the horrendous tragedies they had experienced, was their extraordinary good cheer, optimism, friendliness, and generosity. They had found freedom in exile, but they brought with them human qualities of wisdom and compassion that I valued above all else. I became so drawn to the integrated worldview, values, and way of life presented by the Tibetan scholars and contemplatives with whom I studied that for years I sought total immersion in this culture that was so far removed from my own. Here I found deep spiritual truths similar to those I believed to be embedded in Christianity, but I also encountered a highly intelligent matrix of rational theories and contemplative practices designed to put those theories to the empirical test.

At last I felt I had found a worldview that satisfied my longing for spiritual truths and values, integrated with rational theories and methods of inquiry into the nature and potentials of consciousness and its relation to the natural environment. Eventually, though, it became increasingly obvious to me that in abandoning my native culture and immersing myself in an alien one, I had fragmented myself further. So after fourteen years in exile from the mainstream of Western society, I chose to return to my homeland and to complete my undergraduate education at Amherst College, where I decided to focus my studies on the paradigm of modern science: physics. Thereafter, to further integrate my understanding of science and religion, I earned a doctorate in religious studies at Stanford University, where I studied comparative religion, psychology, and the philosophy of science.

A central concern of all my studies with the Tibetan and Western scholars has been a deep fascination with the nature and potentials of consciousness. This interest has been enormously enriched by my serving as an interpreter and participant in a series of conferences with the Dalai Lama and other Buddhist monks together with various groups of distinguished cognitive scientists, physicists, and philosophers. The first of these "Mind and Life" conferences took place in Dharamsala in 1987, and they have continued on a biannual basis since then. An extraordinary quality of these meetings has been the open-minded yet critical attitude of the Buddhists and the scientists, both eager to expand their horizons by learning of the methods of inquiry and the insights of the other. Published accounts of these meetings have been received with growing interest by people interested in crosscultural and interdisciplinary dialogue, especially concerning the nature of the mind.[12] Such collaboration marks a stark contrast to the more traditional stances of scientists regarding religion as a mere obstacle to discovery and religious people regarding science as a threat to the validity of their creeds. Similar dialogues have been occurring among scientists, philosophers, and members of other religions as recorded in such journals as *Science and Spirit* and the *Journal of Consciousness Studies*.

Another promising trend in recent years has been the growing number of conferences and dialogues among representatives of the world's religions during which the participants seek to enrich their own spiritual practice by learning from the insights of other traditions.[13] I have sought out such encounters myself, including participating in a meditation retreat with the Dalai Lama and a group of Christian contemplatives in Prato, Italy, during the spring of 1999. The rise of nonsectarian interest in the experiential dimensions of contemplative practice is a wonderful departure from the adversarial attitude that has plagued relations among religions for centuries.

These two recent trends have extremely few precedents in human history, and I believe they are the vanguards for devising a truly contemplative science that may shed light—for religious believers and scientists alike—on the nature, origins, and potentials of consciousness. This movement may form the basis for a noetic revolution in which we rediscover not only our early Western roots but our deep global roots, East and West, contemplative and scientific. Rather than regarding science as the one center of our universe, with all other modes of inquiry being peripheral to it, we are now in a position to recognize that other civilizations in the past and the present have their own valid modes of inquiry that may profoundly complement those of modern science. Whenever any institution monopolizes the epistemic authority for a civilization—with all the wealth, power, and prestige that that entails—it is bound to strongly resist anyone who seeks to break that monopoly. But the very health of science requires that it be challenged by ideas and empirical modes of inquiry that are alien to it.

The Scientific Revolution that marked the beginning of the modern era introduced a fresh skepticism regarding deeply cherished, unquestioned assumptions, and it introduced new methods for exploring the natural world. This is precisely the aim of this book, in which I argue for the importance of a plurality of methods and theories to break the domination of any one dogma that insists that the world must be conceived and explored only according to its dictates. This book stands in opposition to all dogmatisms—ranging from the religious to the scientistic—that insist on the acceptance of their doctrines as the sole means of understanding the world and solving human problems.

To understand the relation between science and religion, especially pertaining to the exploration of the nature of consciousness, it is crucial to identify the metaphysical doctrine that underlies and structures virtually all contemporary scientific research. The basic principles of this doctrine, commonly known as scientific materialism—namely, objectivism, monism, universalism, reductionism, the closure principle, and physicalism—are analogous to the axioms of Euclidean geometry. From the fourth century BCE until the nineteenth century, mathematicians commonly assumed that Euclid's axioms were self-evident, absolutely certain truths of the real, physical world; and this view seemed bolstered by the success of physical applications of these principles during the Scientific Revolution. But in 1813, Carl Friedrich Gauss devised a system of geometry that rejected one of Euclid's basic postulates, and in 1830 János Bolyai and Nikolay Lobachevsky proposed that none of those postulates are either true or false of the objective world—they are simply the rules of the game. During the last three decades of the nineteenth century, a variety of geometries were proposed; and mathematicians gradually came over to the view that Euclid's axioms were true for the world of Euclidean geometry but could no longer be construed as absolute truths of the objective, physical universe.

In this book I argue that the fundamental principles of scientific materialism, while true for the world of scientific materialism, are not necessarily true for reality as a whole. These principles have helped us understand a certain range of objective natural phenomena, particularly those described adequately by classical mechanics, and this has led many scientists to believe they are universally valid. But they have simultaneously obscured a wide range of subjective phenomena, including consciousness itself, and in this way dogmatic adherence to these assumptions has limited scientific research and impoverished our understanding of nature as a whole.

Toward the end of the nineteenth century, most scientists implicitly accepted a type of mechanistic materialism that had been wonderfully successful in explaining thermodynamics; they therefore assumed it would also explain electromagnetism. They were simply unable to imagine that the world could contain anything that fell outside the domain of their view of the world. Science proved them wrong. In this book I argue that in an analogous way, rigorous inquiry into the nature of consciousness may upset many of the assumptions of scientific materialism, which has erroneously excluded the subjectively experienced mind from the domain of the natural world.

The rest of this book comprises three parts. In part I I distinguish four elements of the scientific tradition, namely science itself, the philosophical view of scientific realism, the metaphysical doctrine of scientific materialism, and the fundamentalist creed of scientism. After arguing that scientific materialism, unlike science, has taken on the role of a religion, with all its taboos and heresies, I then trace the historical development of this doctrine and its close ties to Christian theology. In part II I present an alternative matrix of theories and practices for exploring consciousness, drawing from the writings of Western scholars and contemplatives such as Augustine, William James, Hilary Putnam, and Robert Forman and Eastern scholars and contemplatives such as Buddhaghosa, Vasubandhu, Asaṅga, and Padmasambhava. The approach outlined in part II for a science of consciousness differs profoundly from the theories and methods of modern cognitive science, so in part III I discuss what I perceive to be the limitations and defects of the scientific study of the mind pursued within the metaphysical framework of scientific materialism. In my conclusion I present guidelines for a contemplative science of the mind that draws from both our global spiritual heritage and our scientific heritage. What is needed, I believe, is a discipline, embracing a range of modes of scientific inquiry into the nature of consciousness, that takes firsthand experience seriously and devises means of exploring it with scientific precision. Such a discipline has the potential to be profoundly contemplative as well as rigorously scientific, and I believe it is the most promising, pluralistic mode of inquiry for discovering deep truths concerning consciousness and its role in the natural world.

PART I

The Ideology of
Scientific Materialism

1

FOUR DIMENSIONS OF
THE SCIENTIFIC TRADITION

Since the Scientific Revolution, claims have been made about science and on behalf of science that include not only scientific but philosophical and theological assertions. When such a wide range of issues is included within the category of scientific knowledge, distinguishing science from a religious-like ideology becomes difficult; and when the authority of science is invoked in support of metaphysical positions, further problems arise. Thus, it is crucial to identify the salient features of distinct aspects of the scientific tradition, namely, science itself, the philosophical view known as scientific realism, the metaphysical ideology of scientific materialism, and the dogmatic form of that ideology known as scientism.

Science

Let me begin by taking note of some of the characteristics of science in its "disembodied" form, that is, divorced from its philosophical and theological underpinnings. Science is a discipline of inquiry entailing rigorous observation and experimentation, followed by rational, often quantitative, analysis; and its theories characteristically make predictions that can be put to the empirical test, in which they may turn out to be wrong, and the theory is thereby invalidated. One of the central ideals of this discipline is that of a disengaged observer, capable of objectifying the surrounding world and suppressing emotions, inclinations, fears, and compulsions in order to pursue research in an unbiased and rational manner.

Another ideal of science is skepticism: one seeks to identify unquestioned assumptions, to question common sense, and to critically examine appearances themselves, for they have often been found to be misleading. The more deeply rooted an assumption or belief is and the more widely it is accepted by one's peers, the more challenging it is to question it, particularly in public. But time and again scientists have risen to this challenge and thereby broadened the scope of human knowledge. On the other hand, if one pushes skepticism to an extreme by excessively doubting one's own and others' beliefs, modes of inquiry, and discoveries, then scientific progress can grind to a halt. Despite the ideal of healthy skepticism in science, there must also be a place for faith, or informed confidence, regarding the advances of earlier generations of scientists and the work of one's contemporaries.

Science frequently involves experimentation, and its theories are aimed at intelligible explanation and predictive ability with respect to those phenomena. Truth in science is determined by the empirical feedback of success in one's predictions, and this pragmatic criterion sets science apart from other disciplines such as philosophy and religion. In this way scientific theories are formulated and tested, with advances in knowledge giving rise to further hypotheses. Scientific hypotheses can be refuted by means of reasoning—for example, if they are found to be internally inconsistent—or by empirical observation or both. However, a hypothesis may also be saved from falsification by modifying surrounding hypotheses or by modifying the interpretation of the empirical data.[1] Although modern science originated with the quest for absolutely certain knowledge of the natural world, scientific knowledge is now usually presented as tentative and subject to change, largely because even some of the most seemingly secure scientific principles have been refuted.

These comments are intentionally very general, so that they pertain to the many disparate branches of science, ranging from theoretical physics to wildlife biology. Given this enormous diversity of fields and methodologies within science, I have chosen to point to an array of salient features of scientific inquiry that provides a basis for recognizing what have been called family resemblances among the sciences. This seems more useful than trying to set down an airtight definition of science as a whole or to identify some "essence" to science that purportedly differentiates it from other modes of inquiry.

While scientific *modes of inquiry* are compatible with many other approaches, including those of philosophy and religion, some aspects of scientific *knowledge* are clearly incompatible with other worldviews, including religious ones. In such cases, the seeker of truth will accept what has clearly been demonstrated to be true by means of rigorous scientific research, while taking care to distinguish such facts from the metaphysically loaded interpretations that scientists may impose upon, and conflate with, those facts. Science has enjoyed enormous successes in explaining objective physical events according to quantitative mathematical laws, but it has been less successful in explaining or predicting subjective mental events. While it has provided humanity with an unprecedented degree of control over the physical world, it has not shown us how to control our own minds from within. And while it has greatly enhanced the physical well-being and security of much of the world's population, it has made little progress in discovering strategies for finding greater personal happiness, mental health, compassion and altruism, or social harmony.

When giving a general overview of science, it is very easy to exaggerate the unity of this body of knowledge, for over the past four hundred years a wide range of scientific methods and theories has been devised, based on diverse metaphysical principles. The most remarkable point here is that many of these divergent approaches have produced empirically acceptable theories. The assumption of a single worldview embracing all of science, however, is misleading, for there have always been many different scientific maps of the natural world, many of which are drawn from different metaphysical and theological viewpoints.[2]

Scientific Realism

Scientific realism is not science but a philosophical interpretation of scientific knowledge and its relation to the world. There are many versions of scientific realism as well as antirealism, the latter variously associated with instrumentalism, constructivism, and empiricism. Advocates of scientific realism believe that the formulation of scientific theories aims to give us a literally true story of what the world is like and that the acceptance of a scientific theory involves the belief that it is true.[3] It is important to note that advocacy of this view does not necessarily imply the assertion that present scientific theories *are* literally true representations of the world; it implies only that they *aim to be*. Simply put, scientific realism asserts that science is trying to discover what is really going on in nature, beyond the scope of appearances; it is aiming at understanding how nature works and why it is the way it is.

In the preceding description of scientific realism, if "the world" is regarded as the universe as it exists in itself, independent of human concepts and language, scientific realism becomes a form of metaphysical realism. In this context, the objects of scientific inquiry are thought to be describable in principle in and of themselves; and they are believed to exist objectively, independently of any descriptions or interpretations imputed upon them by any subjects.[4] Although most of the objects within the objective world are not accessible to the unaided human senses, it is believed that they can nevertheless be discovered, investigated, and described with the empirical and rational tools of science. In the words of philosopher Ernan McMullin, science enlarges our world "through retroductive inference to structures, processes and entities postulated to be causally responsible for the regularities established by the experimental scientist, or for the individual 'traces' with which historical sciences like geology and evolutionary biology are concerned."[5] This retroductive approach has been enormously successful, especially in inferentially discovering causes that are subsequently observed.

Scientific realism has certainly dominated much of scientific thinking throughout history, but it has always had to defend itself against other philosophical perspectives. The scientific realist Galileo, for instance, had to respond to the philosophical arguments of Cardinal Bellarmine, who maintained that scientific theories should be regarded simply as ways of *making appearances intelligible*, without presuming to describe the real nature of the world beyond the veil of appearances.[6] Isaac Newton (1642–1727), the founder of classical mechanics, and James Clerk Maxwell (1831–79), the founder of the modern theory of electromagnetism, are both known for their internal struggles with realist and antirealist philosophical interpretations of scientific knowledge.

Nowadays, most philosophers of physics, the most mature of the sciences, have distanced themselves from scientific realism, adopting views closer to those of Cardinal Bellarmine than of Galileo. Bas van Fraassen, for example, advocates a form of constructive empiricism whose central theme is that science aims to give us intelligible accounts of empirical evidence. That, he maintains, is all that is necessary to accept a scientific theory.[7] The debate goes on, however between various versions of scientific realism and antirealism. Ian Hacking, for instance, mounts a rigorous defense of scientific realism, pointing out that we can be sure of the reality of macro-objects because of what we can do with them and to them, and what they can do to us. Similarly, he argues, if we can measure, manipulate, and understand the causal powers of scientific micro-entities, we have good reason to believe that they are real and theory-independent.[8] Like Hacking, most contemporary philosophers of neuroscience, among the youngest of sciences, adopt scientific realism.[9] Some people go so far as to argue that scientific realism is a psychological prerequisite for successful research in science. Others, such as van Fraassen, maintain that the alternative to reifying the contents of science is to equate science not with a dogma but with a quest in which one immerses oneself in a worldview without reifying it, accepting scientific theories as constructs that are more or less successful in making the appearances of the natural world intelligible.

As scientists devise one mathematical structure after another to describe the world, many of them have faith that this process will eventually yield an ultimate "best theory." Some reify the fundamental laws of physics so far as to attribute qualities to them that are traditionally ascribed only to God, claiming that they are universal, absolute, omnipotent, and eternal, existing independently even of the state of the universe. Mathematical laws of nature, however, while "saving the appearances" with enormous success, do not provide a picture of what the world is like independent of those appearances; and given the wide variety of interpretations of such quantitative laws, it takes a leap of faith to believe that any one of them will prove to be the "right one." Rather, one's choice of constructs—for example, which of the interpretations of quantum mechanics or even Newtonian mechanics one chooses—at any time seems to be largely a matter of temperament; for throughout the history of science multiple theories have commonly accounted equally well for observed phenomena. Scientific realism tends to downplay the role of such subjective influences in scientific research, whereas antirealism tends to isolate the real, objective world from subjectivity, making it unknowable in principle. Throughout this work, the term *subjectivity* refers to all conscious and unconscious *personal* influences, including individual consciousness itself, and all personal, individual goals, attitudes, and points of view. As such, each instance of subjectivity is confined to a specific locality in both time and space.

The tenets of scientific realism, unlike those of science, are not prone to refutation by empirical evidence; however, like other philosophies, they can in principle be refuted by means of rational argument. Scientific realism is compatible with a wide range of religious doctrines, as well as with atheism.

Scientific Materialism

The Essential Principles of Scientific Materialism

According to science, empirical data always have the last word and there is no place for dogmas, sacrosanct theories, or a priori statements. Nevertheless, science has progressed together with the ideology of scientific materialism that does embody a number of sacrosanct theories and a priori statements, namely the principles of objectivism, monism, universalism, reductionism, the closure principle, and physicalism. While these metaphysical principles of scientific materialism are not matters of scientific fact, they are commonly presented in science classrooms, scientific writings, and the popular media as if they were on a par with genuine scientific theories that are subject to empirical verification or refutation. Rarely are these principles taught as a discrete ideology, and many students of science may not be consciously aware of them at all. Scientific knowledge advances in part by a healthy skepticism of long-cherished assumptions, but it is impossible to be skeptical of something of which we are not even conscious.

While scientific materialism actually subsumes science and scientific realism, it is often misleadingly equated with science itself, especially by its advocates. Although the claims of scientific materialism go far beyond the legitimate domains of science, most scientific research is now conducted within the metaphysical framework of this ideology. Thus, it is easy to conflate the two, but in so doing, one overlooks possibilities for scientific research that do not conform to this belief system.

The Scientific Revolution occurred in defiance of the scholastic dogma of its day, and since then, science has advanced by formulating new theories and then revising them or replacing them with better theories as the scope of its empirical knowledge has increased. When old dogmas are challenged, however, it is difficult, for people to resist the temptation to form new dogmas to replace the old ones, for there is something profoundly unsettling about questioning our deepest assumptions. By the term "dogma" I mean a coherent, universally applied worldview consisting of a collection of beliefs and attitudes that call for a person's intellectual and emotional allegiance. A dogma, therefore, has a power over individuals and communities that is far greater than the power of mere facts and fact-related theories. Indeed, a dogma may prevail despite the most obvious contrary evidence, and commitment to a dogma may grow all the more zealous when obstacles are met. Thus, dogmatists often appear to be incapable of learning from any kind of experience that is not authorized by the dictates of their creed.[10] The irrationality of dogmatism has been presented as one of the strongest arguments against all forms of religion, but let me now examine the principles of scientific materialism to see whether science, too, has become constrained by its own unique dogma.

Objectivism. As noted earlier, perhaps the most central ideal of science has been the pure objectification of the natural world, and, implicitly, the exclusion of subjective contamination from the pursuit of scientific knowledge. This ideal has so captured the modern mind that *scientific knowledge* is now often simply equated with *objective knowledge.* The principle of objectivism demands that science deals with empirical facts testable by empirical methods and verifiable by third-person means. This principle has proven to be very useful in revealing a wide range of facts that are equally accessible to all competent observers. Such facts must be public rather than private; that is to say, they must be accessible to more than one observer. However, there are many other empirical facts—most obviously, our own subjective mental events—that are accessible only by first-person means and of which the only competent observer is oneself.

Another aspect of this principle is that scientific knowledge must be epistemically objective, that is, observer independent. In its most defensible guise, this ideal demands that scientists strive to be as free as possible of bias and prejudice in their collection and interpretation of empirical data. In its least defensible form, it demands that scientific knowledge must be free of any subjective, nonscientific influences. This, of course, has never been true of science or any other branch of human inquiry, as has been amply demonstrated in Thomas Kuhn's provocative work *The Structure of Scientific Revolutions.* Even the renowned biologist Jacques Monod, a staunch advocate of scientific materialism, acknowledges that the postulate of objectivity as a condition for true knowledge constitutes what he calls an ethical choice, rather than a matter of fact. This assertion of Monod's implies that this principle is not the result of research but is rather a premise that guides a certain kind of research, while prohibiting other types of research from being conducted.[11]

The principle of objectivism, in the sense of the demand for observer independence, simply cannot accommodate the study of subjective phenomena, for it directs one's attention only to those objects that exist independently of one's own subjective awareness. It is no wonder then that science presents us with a view of a world in which our own subjective existence is not acknowledged and the notion of the meaning of our existence cannot even be raised.

Monism. According to this principle in its scientific guise, there is one unified universe consisting of generally one kind of stuff, which can be described completely by physics. This metaphysical principle is closely conjoined with another belief, known as universalism, which asserts that natural, quantifiable, regular laws govern the course of events in the universe uniformly throughout all of space and time.

For Hellenistic thinkers, phenomena were defined as things, events, and processes that can be seen, in contrast to noumena, which were thought to be things as they are in themselves. According to scientific materialism, however, phenomena have come to be identified as things, events, or processes that occur regularly under definite circumstances. The metaphysical principles that constitute scientific monism have proven to be enormously valuable guidelines for investigating a wide range of phenomena, specifically those that are physical, quantifiable, orderly, and repeatable. On the other hand, they give no account of nonphysical, purely qualitative, sporadic, and unique phenomena. Thus, once again, subjective mental phenomena— which are not demonstrably physical in nature, do not lend themselves directly to quantitative measurement or analysis, frequently appear disorderly, and at times include phenomena that are not evidently repeatable— seem to fall outside the bounds of those principles.

Reductionism. As soon as one accepts the monistic principle that the entire universe consists of only one kind of stuff, namely one that can be described completely by physics, one must identify the nature of this basic stuff. In the twentieth century, many scientists concluded that the world is fundamentally composed of elementary particles of mass/energy. The principle of reductionism augments this view by asserting that macro-phenomena such as the behavior of human cells are the causal results of micro-phenomena (ultimately, the behavior of the elementary particles that constitute the cells). This metaphysical principle is succinctly stated by physicist Earnest Rutherford: "if we knew the constitution of atoms we ought to be able to predict everything that is happening in the universe."[12] Thus, elementary particle physics is thought to deserve the title of the most fundamental description of the world. In short, the metaphysical principle of reductionism declares that there is nothing that living or nonliving things do that cannot be understood from the point of view that they are made of atoms acting according to the laws of physics.[13] Some scientists, however, far more cautiously embrace reductionism as a *method*, which has proven highly useful in many areas of research, without adopting it as a *belief* about the actual nature of reality.[14]

The principle of reductionism has also been applied to the study of psychology and religious experience. The assumption here is that for the more advanced forms of living organisms, behavior and conscious states can be best understood in terms of the more primitive. For example, the pioneering psychologist Wilhelm Wundt advocated a primitive form of internal perception (*innere Wahrnehmung*) to be used as a tool of scientific psychology for understanding the human mind. Confining his research to the study of perception and sensation, he excluded the observation of thought processes, feelings, their complex connections, and the affects and processes of volition. Another trend in late nineteenth-century psychology was the attempt, inspired by the successes of chemistry, to devise a "periodic chart" of the basic elements of mental activity, understood in terms of very brief, primitive events.

Upon the demise of the introspectionist movement in modern psychology in the early years of the twentieth century, behaviorism also adopted the principle of reductionism by studying the behavior of animals as a means to understanding human behavior. Moreover, with the perceived failure of introspection as a means of scientific inquiry, many behaviorists simply reduced all mental activity, including consciousness itself, to objective behavior. That is, subjective, internal mental events were reduced to objective, external events, which science was well accustomed to studying. This was a "lateral" reduction of an anomalous-class of phenomena to a more familiar type of processes.

By the 1960s, when the limitations of behaviorism became increasingly apparent in terms of understanding the mind, much of the emphasis shifted to neuroscientific research, which also laterally reduces subjective mental events to objective brain activity. Despite the vast differences in methodologies and theories within the cognitive sciences over the past century, the principle of reductionism has run through all these disciplines, as if they were all conforming to a pre-established creed. Likewise, when modern scholars have sought to understand various forms of religious experience, including mystical experience, this same trend has been very prevalent.

Reductionism, like the other tenets of scientific materialism, has guided scientists in shedding light on those types of phenomena that can be best understood by examining their elementary components. However, the universality of this assumption is increasingly coming into question in science, for example in the context of chaos theory.[15] In the life sciences, too, close attention to the behavior of elementary particles or even individual cells frequently yields *less* understanding than attention to the more global interactions among systems of cells. Moreover, when it comes to scientifically inquiring into the nature and origins of consciousness and other mental events, the principle of reductionism may actually obscure the phenomena one is trying to investigate.

The Closure Principle. The adoption of the principle of reductionism *as it was formulated in twentieth-century scientific materialism* implies another of the principles of this metaphysical dogma: what has come to be known as the closure principle. According to this belief, the physical world is "causally closed"—that is, there are no causal influences on physical events besides other physical events.

The closure principle has proved to be a useful hypothesis for the investigation of a wide range of interactions among physical phenomena; but if there are any nonphysical influences on physical events, unquestioning acceptance of this belief will ensure that those influences will not be recognized. Some scientific materialists have misleadingly argued that the closure principle must be a universal truth because scientific research has found no evidence of any nonphysical influences in the natural world.[16] The distinguished biologist Edward O. Wilson, for instance, declares that the religious belief in a God who directs organic evolution and intervenes in human affairs "is increasingly contravened by biology and the brain sciences."[17]

Natural philosophy, as it was envisioned by Descartes and other participants in the Scientific Revolution, had only the physical world as its proper domain, and this has been largely true of science ever since. Moreover, never in the history of modern science have instruments or methods been devised to detect the presence of nonphysical influences of any kind. Research in modern biology and the brain sciences is conducted with the assumption, hardly ever questioned, that there are no nonphysical influences in organic evolution or in human affairs. So the fact that scientists have not discovered any such influences should hardly come as a surprise. And at this point in history, it is certainly premature to declare that scientific knowledge of organic evolution and brain activity is so complete that nonphysical influences can be absolutely ruled out on purely empirical grounds.

Particularly with regard to the human mind, the closure principle seems to be incompatible with experience, for our conscious mental processes, which have not been demonstrated to be composed of configurations of mass and energy, certainly do appear to influence human behavior. Advocates of the closure principle assume that the *apparent* influence of our desires, beliefs, and intentions on our behavior is actually an illusion—all behavior is in fact determined solely by the interaction of the nervous system with the rest of the body and the physical environment. However, contemporary neuroscience does not even remotely possess sufficient understanding of the brain to verify this assumption on the grounds of empirical evidence. If for no other reason, the fact that modern science does not know how or why consciousness first appeared in terms of the evolution of life on our planet or in the development of a human embryo should make it abundantly clear that the closure principle is a metaphysical belief and not a scientific fact.

Physicalism. With the widespread adoption of reductionism and the closure principle in the nineteenth century, due in part to the widespread acceptance of the principle of the conservation of energy,[18] scientific materialism abandoned its Judeo-Christian origins. No longer could this metaphysical dogma conform to the Judeo-Christian belief in a nonphysical, personal God who intervenes in the course of nature and human history and who responds to the prayers of individuals. By the nineteenth century, the only religion with which scientific materialism remained compatible was Deism, a religion contrived in part by the proponents of scientific materialism itself.

Albert Einstein was among the most eminent scientists educated in the nineteenth century to declare that the concept of a personal God is utterly incompatible with science and that it is the major source of conflict between religion and science. This theological stance, however, did not prevent him from believing in a universal Superior Mind that reveals itself in the world of experience. This Deist view retains the Christian belief that God possesses an absolute perspective on reality, but it denies that God influences natural events in any way.[19] In other words, God is conceived of as an ideal scientist, a purely objective observer who sees reality as it is without any personal, subjective intervention.

Twentieth-century scientific materialism abandoned belief in any form of theism by adopting the principle of physicalism, which states that in reality only physical objects and processes exist. In other words, only configurations of space and of mass/energy and its functions, properties, and emergent epiphenomena are real. A closely related principle maintains that everything that exists is quantifiable, including the individual elements of physical reality, as well as the laws that govern their interactions. At this point scientific materialism becomes compatible only with some of the more primitive nature religions. The "God's-eye view" of reality that was the earlier ideal of scientific materialism has been replaced by the ideal of the "view from nowhere."[20] That is, the ideal of pure objectivity has been retained, but it has been divorced from the theological underpinnings that originally gave it credibility, meaning, and value. Thus, the quasimystical quest of earlier scientists to view God's creation from God's own perspective has been replaced by the ideal of viewing a mindless, meaningless universe from a nonhuman, purely objective perspective.

There are many scientists and philosophers, of course, who deny that physicalism is simply a metaphysical principle. Philosopher Patricia Churchland, for instance, claims that the rejection of consciousness (or any other "spooky stuff" such as a soul or spirit) existing apart from the brain "is a highly probable hypothesis, based on the evidence currently available from physics, chemistry, neuroscience and evolutionary biology."[21] She declares that the assertion of physicalism is an empirical matter, not a question of conceptual analysis, a priori insight, or religious faith. Philosopher Güven Güzeldere asserts in a similar vein that "... contemporary science tells us that the world is made up of nothing over and above 'physical' elements, whatever their nature (waves, particles, etc.)."[22]

Let us assume for the moment that these physicalists are right in asserting that scientists have empirically demonstrated that only physical things and events exist. This would imply that this assertion belongs together with a wide range of other scientific facts—such as the convertibility of mass and energy, the atomic weight of hydrogen, and the nature of photosynthesis—about which there is a very high degree of consensus among the scientific community. Churchland acknowledges that not all philosophers agree with her physicalist belief, but it must also be acknowledged that a very sizable proportion of the scientific community doesn't either. Given that 40 percent of American scientists today believe in a personal God to whom they can pray and that this figure has not changed significantly over the past century, it would seem that if the physicalist hypothesis has been proven empirically during the twentieth century, virtually half of the scientific community in the United States still refuses to acknowledge it. If this is the case, are they prevented from seeing this empirical truth as a result of their commitment to a theistic ideology? If so, this raises a profound qualm about the reliability of the scientific community as a whole to distinguish empirical facts from ideological commitments. One might just as well ask the same question of those scientists who believe that the empirical evidence does confirm the hypothesis of physicalism: Are they overinterpreting scientific evidence to make it conform to their metaphysical beliefs? If we are to trust the scientific community to give unbiased reports of its research, then physicalism *must* be regarded as a matter of conceptual analysis, a priori insight, or religious faith. For there is clearly no scientific consensus on this matter, or even a historical convergence toward such a consensus among scientists.

The Marginalization of the Mind

While the nineteenth-century adoption of the closure principle denied causal efficacy to anything that is nonphysical, the twentieth-century version of physicalism denies that anything nonphysical even exists in reality. This shift has major implications for the relation between the mind and the physical universe. It is noteworthy that, while physical science was well established by the late seventeenth century, a science of the mind was not initiated until a full two centuries later. And even then, particularly in the Anglo-American world, the focus of academic psychology swiftly shifted away from the mind and toward behavior, and then to neuroscience. Only in the latter half of the twentieth century did cognitive psychology, for example, begin to reconsider the functions of the mind as it is experienced firsthand. In the historical development of modern science, the study of the mind occurs only as an afterthought, subsequent to the elaborate development of physics, chemistry, and biology; so it is no coincidence that in the world as conceived by science, the role of the mind in nature has been marginalized. According to this view, the universe is conceived as a giant computer, and the emergence of consciousness during the course of cosmic evolution is attributed solely to the laws of physics, which over the immensity of time give rise to a nearly infinite complexity that is purportedly sufficient to give rise to consciousness. This "explanation" places an enormously heavy explanatory burden on the term "complexity," which in fact explains nothing.

Since the Scientific Revolution, subjectively experienced mental events have gradually lost their status as real entities. Advocates of scientific materialism now variously regard them as mere epiphenomena, as propensities for behavior, as being equivalent to brain activity, or as bearing no existence whatsoever. As one indicator of this phenomenon, it is worth noting the types of discoveries for which Nobel awards have been granted in the fields of physiology and medicine. While it is well known that many mental phenomena—including hope and fear, happiness and depression, trust and suspicion, and belief and disbelief—have profound influences on the human body and state of health, since Nobel awards were first granted in 1901 for discoveries in physiology or medicine, *not a single one has been given for discoveries about the nature of the mind.* One could rationalize this fact by claiming that research into the nature of the mind and its possible influences on the body is not included in the domain of "hard science" and is therefore unworthy of such a distinguished award. But "hard science" in this context means nothing more than science that rigidly conforms to the metaphysical dictates of scientific materialism, even at the cost of ignoring significant aspects of health and disease.

How did the mind, which exerts such a powerful influence in our daily lives and which makes science possible, become so marginalized? In his classic work *The Principles of Psychology,* the American psychologist and philosopher William James (1842–1910) presents a thesis that sheds brilliant light on this issue:

> The subjects adhered to become real subjects, attributes adhered to real attributes, the existence adhered to real existence; whilst the subjects disregarded become imaginary subjects, the attributes disregarded erroneous attributes, and the existence disregarded an existence in no man's land, in the limbo "where footless fancies dwell." . . . Habitually and practically we do not *count* these disregarded things as existents at all . . . they are not even treated as appearances; they are treated as if they were mere waste, equivalent to nothing at all.[23]

James sums up this idea with the assertion that "our *belief and attention are the same fact. For the moment, what we attend to is reality*. . ."[24] A historical illustration of this theme is to be found in the history of behaviorism. In 1913, the American behaviorist John B. Watson wrote that "the time has come when psychology must discard all reference to consciousness,"[25] and he attributed belief in the existence of consciousness to ancient superstitions and magic.[26] Fifteen years later, he expanded this principle by declaring that behaviorists must exclude from their scientific vocabulary "all subjective terms such as sensation, perception, image, desire, purpose, and even thinking and emotion as they are subjectively defined."[27] Behaviorism duly followed this dictum, with the result that in 1953, B. F. Skinner concluded that *mind* and *ideas* are nonexistent entities "invented for the sole purpose of providing spurious explanations. . . . Since mental or psychic events are asserted to lack the dimensions of physical science, we have an additional reason for rejecting them."[28] Assertions concerning subjective experience were similarly denied by certain philosophers of the same period who argued against the very existence of subjective statements.[29]

A similar denial of mental phenomena (Skinner eventually retracted his)[30] is to be found nowadays in a contemporary philosophical school known as eliminative materialism. Proponents of this view, for example, Paul Churchland and Stephen Stich, argue that subjectively experienced mental states should be regarded as nonexistent, on the grounds that the descriptions of such states are irreducible to the language of neuroscience.[31]

Since the time of Galileo, scientific materialism has been absorbed in extraspection: it has focused its attention even beyond the external world of human experience to the objective reality that purportedly lies behind the veil of appearances. This, it deems, is the world of science, and it alone is real; whereas mental phenomena, which are purportedly accessible to introspection, have come to be treated by the advocates of scientific materialism as "mere waste, equivalent to nothing at all."

The central aim of science is to understand and control the objective, physical world; yet the subjective mind, with its powers of observation and reasoning, is, awkwardly, the fundamental instrument of scientific inquiry. With their ideal of absolute objectivity, in which all subjective influences are excluded, the advocates of scientific materialism have sought to exclude the subjective elements of even the human mind. According to this ideal, scientific research is to be conducted in an utterly dispassionate manner, free of all personal biases; and even scientific thinking is portrayed as an impersonal activity. Moreover, instead of human logic and language, scientists are to employ as much as possible the laws of mathematics, which are thought by many to be purely objective rules.

The disdain of scientific materialism for subjectivity has also shaped the very concept of scientific observation. While nonscientific kinds of observation also detect phenomena—such as our joys and sorrows, hopes and fears, ideas and inspirations—they are thought to be tainted by human subjectivity and are therefore suspect. From the perspective of scientific materialism, human sensory perception may be deemed not only unreliable but irrelevant. For a scientific observation to take place, all that is required is a detector, or receptor. The human eye is one type of receptor, which detects a certain range of electromagnetic frequencies, but other instruments also measure this and other types of information, and they are regarded as more reliable.

In common parlance, for an observation to take place, the received information must be transformed into humanly accessible information that is, sooner or later, perceived and understood by a human being. But according to scientific materialism, *observation* is assimilated into the general category of *interactions*, thereby freeing it from the subjectivity of its normal associations. This interpretation is said to be central to grasping what is involved in scientific objectivity in the search for knowledge and the justification of belief.[32]

If we were to accept the assertion of scientific materialism that observation and measurement occur without any relation to consciousness, there would be no valid reason to exclude *any* physical interaction from this category. Not only instruments of artificial intelligence but all phenomena with spatial dimension would always be detecting—that is, observing and measuring—all the phenomena with which they come in contact. Likewise, not only clocks but all physical phenomena that endure in time would be observing the duration of the phenomena with which they come in contact. In other words, every animate and inanimate phenomenon in the entire universe would constantly be observing its spatial and temporal environment.

However, arriving at panpsychism by such a route blurs any real distinction between the statements that everything is conscious and that nothing is conscious. Moreover, from this vantage point it becomes impossible to ascertain the real difference between conscious and unconscious measurements. Thus, the assertion of unconscious observation and measurement has the effect of obscuring the unique, experienced nature of consciousness, which has been ignored by scientific materialism all along.

The Religious Status of Scientific Materialism

The sheer fact that scientific materialism as it was formulated in the twentieth century is incompatible with all the traditional world religions is enough to provoke the question of whether this doctrine has itself become a kind of modern religion. If the only thing that can displace or substitute for one religion is another religion, scientific materialism would appear to fill that role; and there is no question that this dogma has won many converts from traditional religions.

For the advocates of scientific materialism, traditional religions no longer make sense of the world and human existence in light of modern scientific knowledge. In this light, science, for scientific materialists, becomes an indispensable quest for intelligibility, without which the world and human existence become meaningless. But science alone is incapable of grappling with normative and intrinsic values; it cannot point to the purpose of human existence; and the shifting sands of scientific theories do not provide a firm ground from which to view the world. In short, science itself is not a religion, and it cannot serve the functions in human life that a religion must fulfill.

The metaphysical doctrine of scientific materialism, on the other hand, does fulfill these needs for its proponents. Not only does it present a framework within which to live, it provides its followers with a sense of meaning and thereby connects their lives with a greater reality. This meaning and greater reality are included in the concept of development of science and technology aimed at a complete scientific understanding and technological conquest of nature.

Scientific materialists might disagree with this thesis on the grounds that rational, empirical *truths* may refute religious creeds. It is certainly true that scientific research has revealed truths about the natural world that are incompatible with the descriptions of nature found in many prescientific religious doctrines. However, the notion that the principles of scientific materialism, unlike traditional religious beliefs, are evidently true to all open-minded, intelligent people is nothing more than propaganda. To many people who accept these principles, they do indeed seem self-evident and irrefutable, just as the fundamental premises of the world's traditional religions seem self-evident to their most devoted followers. But to the outsider the "truths" of all these creeds may seem nothing more than articles of faith.

To illustrate the dogma-to-dogma confrontation between traditional religions and scientific materialism, let us examine Edward O. Wilson's account of the sources of religion. Wilson's central claim is that religion is instinctive, meaning "only that its sources run deeper than ordinary habit and are in fact hereditary, urged into existence through biases in mental development that are encoded in the genes."[33] He elaborates on this point by drawing a radical distinction between the origins of religion and biology.

> The human mind evolved to believe in gods. It did not evolve to believe in biology. Acceptance of the supernatural conveyed a great advantage throughout prehistory, when the brain was evolving. Thus it is in sharp contrast to the science of biology, which was developed as a product of the modern age and is not underwritten by genetic algorithms. The uncomfortable truth is that the two beliefs are not factually compatible. As a result, those who hunger for both intellectual and religious truth face disquieting choices.[34]

This theory of the origins of religion is a direct product not of any universally compelling scientific evidence but of the principles of scientific materialism. While advocates of this dogma will probably find his explanation plausible and comforting, to believers of more traditional religions it may seem unsubstantiated and offensive. From their perspective, Wilson's speculations may sound more like an evangelical tract condemning the unholy origins of other faiths rather than an unbiased, scientific theory supported by compelling evidence.

Wilson throws to the winds any notion of reconciliation between science and religion, claiming that "[s]cience has always defeated religious dogma point by point when differences between the two were meticulously assessed."[35] When surveying the history of scientific discoveries in the face of religious dogmas, one finds much to support his position. On the other hand, when it comes to ostensibly scientific responses to a wide array of human experiences that do not conform to the metaphysical principles of scientific materialism, one finds them classified as "anomalies" and "mere coincidences." To scientific materialists, such responses may be satisfactory, but to those not of their faith such "explanations" appear inadequate and at times even irrational.

To draw a further parallel between the origins of traditional religions and of scientific materialism, I return to Wilson's own account.

> Successful religions typically begin as cults, which then increase in power and inclusiveness until they achieve tolerance outside the circle of believers. At the core of each religion is a creation myth, which explains how the world began and how the chosen people—those subscribing to the belief system—arrived at its center. Often the mystery, a set of secret instructions and formulas, is available to members who have worked their way to a higher state of enlightenment.[36]

As I will show in the next chapter, the articles of faith of scientific materialism are largely rooted in the philosophical and religious beliefs of ancient Greece and of Judaism and Christianity. During the rise of modern science, the percentage of scientists and their followers who advocated the principles of this new creed were a small minority, or "cult," to use Wilson's term, but by the twentieth century, they had increased in power and inclusiveness until they achieved a tolerance outside their circle of believers. At the core of this creed is an account of cosmogony and evolution, which is based on scientific research that is conducted in conformity with the principles of scientific materialism. Traditional nature religions posit that the nature of our existence in the world is determined by forces and agents that only the priests have access to and can manipulate. In scientific materialism, scientists and engineers have assumed the earlier role of the priests and sorcerers who know and control the mysterious forces of nature.

Scientific materialists are committed to the tradition of science and characteristically display considerable confidence in the authority of science and in its future progress. The noble ideal of this doctrine is that the march of science will proceed to an increasingly complete and flawless understanding of the universe and through the resultant control of the natural world will provide solutions to humanity's problems. The most optimistic of these proponents go so far as to suggest that scientific knowledge of the physical world will be essentially complete in the near future.[37]

Modern science was originally conceived of as the pursuit of absolute, certain knowledge of the natural world, as this ideal is expressed in the writings of Galileo and Newton. However, as science matured, many scientists have relinquished their claims to absolute, certain truth, as old scientific "truths" have been successively modified or abandoned and replaced by new theories. In this regard, a religious creed may be said to differ from a scientific theory in claiming to embody eternal and absolutely certain truth, while science is always tentative and open to eventual modifications in its present theories. Astute scientists are aware that their methods are incapable of arriving at a complete and final demonstration. Scientific materialists, in contrast, tend to hold onto their metaphysical principles with all the tenacity of religious believers. Just as medieval theology took the most general principles as its starting point, so did scientific materialism begin with large metaphysical assumptions and not simply with particular facts discovered by observation or experiment.

The origins of scientific materialism are permeated with theological beliefs; this doctrine was founded with ideals that were largely religious in nature; and it has traditionally been defended on theological grounds. In addition, this creed has drawn converts from other religions, and it attempts to fulfill the religious needs of its followers; furthermore, like many other religions, it demands exclusive allegiance. It is therefore misleading for its devotees to present it as an antithesis of religion, when in reality it is a modern kind of nature religion.

The Central Totem and Taboo of Scientific Materialism

In this consideration of scientific materialism, in contrast to science, as a religious creed, let us examine the theory of religion presented by the French sociologist Emile Durkheim (1858–1917). According to this pioneer in the sociology of religion, religious beliefs are representations that express the nature of sacred things and the relations they sustain, either with each other or with profane things.[38] Although it was far from his intent to apply this concept to scientific materialism—indeed, he sought to replace religion with science—I shall argue that scientific materialism meets the criteria of a religion according to Durkheim's theory.

In Durkheim's view, in religious belief sacred things constitute an ideal world that makes intelligible the profane world of the senses, while bearing a greater significance and reality than mundane things. Of crucial importance is the fact that the sacred influences the profane, but the profane should never touch—and thereby contaminate—the sacred.[39] This separation of the sacred from the profane gives rise to the formulation of taboos, or interdictions. While it may be ontologically impossible for the mundane to touch the sacred, on a practical level it is certainly possible to contaminate one's *experience of the sacred* due to influences of the profane. And this must be avoided at all costs.

According to Durkheim, "[t]here is no religion where there are no interdictions and where they do not play a considerable part."[40] Moreover, the most important and extended type of religious taboo is

> the one which separates . . . all that is sacred from all that is profane. So it is derived immediately from the notion of sacredness itself, and it limits itself to expressing and realizing this. Thus it furnishes the material for a veritable cult, and even of a cult which is at the basis of all the others; for the attitude which it prescribes is one from which the worshipper must never depart in all his relations with the sacred. It is what we call the negative cult. We may say that its interdicts are the religious interdicts *par excellence*.[41]

Violation of such taboos is not only thought to result in misfortune for the guilty person due to the natural order of things but also calls for punishment by humans, for it offends public opinion, which retaliates against it.

Traditionally, human communities gain access to the sacred, or ideal, world by means of religious beliefs and practices. As Durkheim develops the main theme of his classic work *The Elementary Forms of the Religious Life*, he addresses the issue of *mana*, a transpersonal, universal force that is central to all religions. This alone is the real object of any religious cult, and its chief representation is the totem. "The totem is the means by which an individual is put into relations with this source of energy"[42] and is the source of the moral life of the clan. Finally, it is the totem that provides a clan with its unique sense of identity. Concerning the relations among diverse totemic groups, Durkheim writes:

> each totemic group is only a chapel of the tribal Church; but it is a chapel enjoying a large independence. The cult celebrated there, though not a self-sufficing whole, has only external relations with the others; they are juxtaposed without interpenetrating; the totem of a clan is only fully sacred for that clan.... The idea of a single and universal *mana* could be born only at the moment when a tribal religion developed over and above the clan cults and absorbed them more or less completely. It is with the sense of tribal unity that there awakens the sense of the substantial unity of the world.[43]

During the formative years of the Scientific Revolution, a number of eminent scientists, for example, Robert Boyle (1627–91), regarded scientific inquiry as a form of worship performed by scientists in the temple of nature. The activity of science was sacred because it sought to understand God's Creation and, thereby, to draw closer to the mind of God. But as science progressed, first God's role as ruler of creation and then his role as creator and sustainer of nature came to be challenged. In the eventual absence of the divine, only the temple of nature remained, under the watchful eyes of its scientist-priests. In this way the objective world of nature has come to take the place of the sacred. This is not to say that many scientists actually regard nature as a sacred realm; rather, *the objective world is all that is left to take its place*. According to scientific materialism, it is the objective world and not God that makes intelligible the profane world of a sense appearances, which is thoroughly tainted by subjectivity.

Numerous scientists of the seventeenth century, from Galileo to Newton, affirmed the Cartesian dualism of the primary properties of the physical world versus the secondary properties associated with human perception. The goal of science was to see beyond the veil of these secondary properties to the true nature of the physical world. This is not to say that many scientists were not of two (or more) minds on this matter. Newton, for example, declared in his *Mathematical Principles* that he refused to make any hypotheses about the underlying mechanism of gravity as it exists apart from phenomena. However, in his *Opticks* he did succumb to the powerful urge to theorize about the inherent nature of gravity.

The scientists of this era were also deeply intent on transcending the theological disputes that gave rise to sectarian rivalry; and these disputes they attributed to the fallibility of human, subjective interpretations. Galileo argued that the truths of nature are inexorable and immutable and no truths that the physical world sets before us ought to be called into question.[44] Reasoning, in his view, was certainly necessary; but the type that was needed was based on mathematics, which God placed in nature, rather than mere human reasoning, which led to multiple interpretations and disputes. Einstein embraced the same theme when he commented that nature, in its own right, can be captured not in any human language but with pure mathematical thought alone, whose propositions are absolutely certain and indisputable.[45]

Thus, within the context of scientific materialism, the subjective realm of human perception, reasoning, and language are set in opposition to the objective realm of the physical world, its inexorable laws, and mathematics. While the objective realm has taken the place of the sacred, the subjective realm has taken the place of the profane.

According to this view, objective reality thoroughly conditions subjective processes. Thus, the brain, one's genetic constitution, and other external, objective stimuli determine mental, emotional, and sensory functions; but apart from those objective influences, no subjectively experienced events, as such, exert any causal influence in the real world. Scientific materialism maintains that brain functions, to which subjective experience is reduced, interact extensively with the external environment. However, this reduction of conscious experience to brain states simply reinforces the preceding point: subjective experience can be allowed to influence the objective world only insofar as such experience is reduced to objective processes.

Durkheim asserts that the concept of mana is the precursor of the scientific concept of energy that was developed during the nineteenth century. Its essential characteristic is that it is located nowhere definitely yet is everywhere present, manifesting in a myriad of diverse forms. According to Durkheim, mana is seen as the objective reality that underlies, empowers, and regulates all physical phenomena. It is altogether distinct from physical power and is in a way supernatural, but it shows itself in physical force or any kind of power or excellence a person possesses. In short, "[a]ll forms of life and all the effects of the action, either of men or of living beings or of simple minerals, are attributed to its influence."[46] In primitive religions the notion of mana served to explain "the world of experienced realities,"[47] which, for Durkheim, were social realities.

One chief distinction between religious notions of mana and the scientific concept of mass/energy is that the latter is regarded as purely physical, whereas the former is not. Note, however, that according to physicist Richard Feynman, a staunch scientific materialist, the conservation of energy is a mathematical principle, not a description of a mechanism or anything concrete. "It is important to realize that in physics today," he writes, "we have no knowledge of what energy *is*."[48] There is certainly no consensus among physicists that energy is some physical stuff existing in the objective world, but if it is not, it is even less clear exactly what it is.[49] Nevertheless, like mana, it is still thought to underlie, empower, and regulate all physical phenomena, and it manifests in physical force.

If the sole preoccupation of science is understanding and controlling forms of mass/energy, what means does science employ to gain access to this objective reality? Science's totem, I suggest, is the scientific method, "the means by which an individual is put into relations with this source of energy." The term *scientific method* is every bit as multifaceted as is *totem* in the context of primitive religions. For each "clan" within the scientific community—from elementary particle physicists to ecologists—the totem of the scientific method appears under different guises. The scientific method in the abstract is associated with careful observation and experimentation, inductive reasoning, and quantitative analysis. But for specific clans of scientists, certain of these characteristics are marginal, while others are dominant.

Mathematical physicists, for example, may hardly concern themselves with observation or experimentation; and wildlife biologists may at times neglect quantitative analysis. Despite the profound differences in the scientific method for diverse branches of science, the *ideal* of this means of inquiry is the source of ideals and the very identity of the scientific community as a whole. Specific versions of the scientific method further distinguish different branches of science and stand out as the totem for each one in the event of interscientific disputes. Yet despite the element of discord within the scientific community, there is widespread unity in the sense that all scientists seek to comprehend phenomena in terms of the objective world. Finally, and perhaps most important, the scientist's methodology must itself be *objective*, that is, as free as possible from all subjective influences.

Prior to the nineteenth century, diverse sciences each enjoyed a large degree of independence. For example, atomic theories were developed quite autonomously in the fields of chemistry and physics. As in Durkheim's description of totemic groups, the sciences were juxtaposed without extensive interpenetration. However, with the development of the principle of energy conservation in the nineteenth century, a single and universal concept of energy was conceived, and it exerted a powerful unifying influence on the sciences.[50] Thus, as in Durkheim's account, with this sense of "tribal unity" there awakened a sense of the substantial unity of the world.[51]

These applications to modern science of Durkheim's thoughts on the sacred versus the profane, mana, and the totem conform closely to his claim that science pursues the same end as religion and is better fitted to it. In his view, scientific thought, which he claims is "only a more perfect form of religious thought,"[52] properly supplants the cognitive authority of religion altogether.

From a traditional religious perspective it may seem that science has banished the sacred and left us with a world that is bleakly and utterly profane. From the perspective of many scientists, however, a religious orientation is not at all alien to science. A religious status has been attributed to science at least since the seventeenth century, as we find in the writings of Galileo, Boyle, and Newton. Moreover, beginning in the late eighteenth century, scientific materialists began to speak of their scientific awakening in terms that might be used of a religious conversion. To take but one example, Lyon Playfair, one of the most energetic evangelists of scientific materialism in nineteenth-century Britain, declared in 1853 that "science is a religion and its philosophers are the priests of nature."[53] A century later Albert Einstein was to claim that "you will hardly find one among the profounder sort of scientific minds without a religious feeling of his own" and that "in this materialistic age of ours the serious scientific workers are the only profoundly religious people."[54]

Einstein's own conversion to this new faith is an interesting case in point. Despite the fact that he was the son of entirely irreligious Jewish parents, as a child he was deeply drawn to the faith of his forefathers. This shifted, however, at the age of twelve, when his encounter with popular scientific books led him to the conclusion that many of the biblical accounts could not be true. This resulted in his conversion from Judaism to scientific materialism.[55] Einstein did not apparently draw any clear distinction between science and scientific materialism but, like Durkheim, conflated the two. Whether or not it is legitimate to distinguish them as sharply as I have argued here, religious attitudes within the scientific tradition are neither new nor uncommon; indeed, they appear to run throughout most of the history of modern science.

Scientism

Although signs of scientism can be found in writings as early as the seventeenth century, they have become far more prevalent since the nineteenth century with the rise of scientific positivism, a view that originated with Auguste Comte. Its three central assertions are that (1) science is our only source of genuine knowledge about the world, (2) science is the only way to understand humanity's place in the world, and (3) science provides the only credible view of the world as a whole. Scientism subsumes scientific materialism (and, thus, scientific realism and science), but it is normally equated by its proponents with science itself. The term "scientism" is invariably used in a pejorative sense, so even those who accept the above three tenets of this doctrine do not call themselves advocates of scientism. They simply say they believe wholeheartedly in science.

Scientism has been depicted in various ways by its detractors. It has been described as the doctrine that science knows or will soon know all the answers and has been said to judge disbelief in its own assertions as a sign of ignorance or stupidity. Scientism unjustifiably extends the authority of science beyond its proper limits, and it assumes that science can solve all of humanity's problems. Expressions of scientism appear in science textbooks, the popular scientific press, and professional scientific literature. It has made deep inroads into the humanities, and its unexamined assumptions have a hold within nearly every field of scholarship.[56]

In short, scientism adopts an absolutist perspective on reality and denies the value of all other avenues of inquiry and knowledge. Much as the fundamentalists of traditional religions regard the revealed message of their scriptures as self-evident, requiring little or no interpretation on the part of humanity, so do advocates of scientism regard the Book of Nature as revealing its own truths to objective, impersonal observation and reasoning. According to this view, there are no significant philosophical problems in the scientific acquisition of knowledge, and the subjective cogitations on this subject by philosophers are largely useless.

Scientism presents the body of scientific knowledge as a unified whole, just as nature is a unified whole, for the former is regarded as a steadily improving representation of the latter. Thus, the notion of relativity of perspective and methodology in the investigation of nature is seen as wholly spurious. Much as religious fundamentalism presents only an idealized caricature of the history of its own beliefs, so does scientism present the history of science as a unswerving march toward Truth, in which earlier errors are systematically replaced with facts. As theistic fundamentalists view the history of their tradition as being guided by the hand of God, so do the proponents of scientism see the history of science as being led by the hand of Nature. In both cases, human influences in the form of personal biases, social values, economic considerations, accidents, and so on are consciously or unconsciously concealed.

Taking into account the role of human subjectivity appears to be equally taboo in both religious and scientistic fundamentalism. According to many schools of religious fundamentalism, the subjective minds of humans are seen as insignificant in relation to the supreme mind of God; and the deeds of humans pale in contrast to the works of the Almighty. Moreover, religious fundamentalists throughout the world tend to overlook the human role in the selection and transmission of their sacred writings, preferring instead to view their scriptures as being the direct expression of the divine. From this utterly objective, transcendent source, devout believers think of themselves as receiving and conveying to others this divinely inspired knowledge, without contamination by their own human subjectivity. In this way, the role of diverse, mutually incompatible human interpretations is minimalized, and the religious doctrine is treated as being a complete, integrated, internally consistent whole, whose divine origin transcends human subjectivity. Thus, the doctrine is presented as being the only viable source of solutions to all the major problems of humanity, and the appropriate response on the part of true believers is to accept its assertions without question and to follow its dictates in a spirit of submission and obedience.

Similarly, according to the dictates of scientism, with regard to scientific observation and quantitative analysis, humans are best replaced by mechanical instruments of detection and computation; and in terms of participation in the act of experimentation, the more passive the human role, the better. Advocates of scientism commonly overlook the subjective, human role of choosing which natural phenomena to investigate, the means of investigating them, and the diversity of human interpretations of research data. Science is presented, like a religious doctrine, as being essentially a complete, integrated, internally consistent whole whose origin in nature transcends human subjectivity. While science provides all genuine knowledge of existence, technology holds all the keys to solving the problems of humanity: environmental, economic, medical, psychological, and social. The appropriate response on the part of the lay public is to be supportive of the scientific community and gratefully receive its technological blessings.

Religious fundamentalists regard those who reject their dogma as being victims of their own sin, especially the sin of pride. Similarly, champions of scientism condemn dissenters from their view as having abandoned reason, for it is inconceivable to them that anyone could be rational and knowledgeable of science yet deny their most cherished scientistic beliefs. In short, scientism is to scientific materialism what fundamentalism is to all traditional religions.

In this chapter I have tried to identify the salient characteristics of scientific materialism and its taboos within the fourfold typology of science, scientific realism, scientific materialism, and scientism. The mingling of religious beliefs and scientific knowledge is not new to the twentieth century; rather, it has characterized the development of modern science all along. To understand this trend more clearly, let us turn now to the history of the interaction between theology and science in the West.

2

THEOLOGICAL IMPULSES IN THE SCIENTIFIC REVOLUTION

The Initial Conception

The seeds of scientific materialism can be found in early Hebrew and Greek religious and philosophical beliefs dating from the sixth century BCE and possibly earlier. Scientific realism, which is an integral philosophical premise of scientific materialism, was profoundly influenced by the biblical assertion that God created the rest of the universe before he created humans. The immediate implication of this belief is that the world experienced by humans exists prior to and independently of the human mind. Thus, the inheritors of this view naturally assume that there is a real, objective world out there; and with the biblical assertion that man is created in the image of God (who is also regarded as being male), there are theological grounds for believing that the mind of man may fathom the universe created by God. In this way the theological grounds of scientific realism were laid.

In order for man to comprehend God's creation, he must divest his modes of inquiry of all that is merely human, which, after all, came at the very end of creation. Man must explore the universe in ways that approximate God's own perspective on creation. He must seek to view the world beyond the confines of his own subjectivity, just as God transcends the natural world. In short, he must seek a purely *objective* (divine) God's-eye view and banish all *subjective* (profane) influences from his empirical and analytical research into the objective universe. In this way the seeds of objectivism were introduced into Mediterranean thought by Jewish, Christian, and Muslim theology.

On the eastern shores of the Mediterranean, the early Ionian thinkers of the sixth century BCE widely assumed that the world is made out of one

41

kind of stuff, commonly thought to be an amorphous, formless kind of matter (*hyle*). Democritus elaborated on this theme by suggesting that the world consists solely of atoms and space. Thus, in the infancy of Western metaphysical speculation, the principles of monism, physicalism, and reductionism were already prevalent. When it came to probing the nature of subjective mental events or consciousness itself, these early philosophers had little to say, in comparison, for instance, to the contemplatives and philosophers of India during that same era. Hippocrates, however, did hypothesize that all mental phenomena are located in the brain, long before there was any compelling empirical evidence to support such a theory.

Thales, Parmenides, and Aristotle all taught that it is impossible for something to arise from nothing, and until the rise of Christianity there was apparently no Greek, Roman, or Jewish Hellenistic thinker who asserted that the world was created from nothing. The first to propose that God created the world ex nihilo were Christian theologians writing in the second century with the intent to refute certain Gnostic and Greek theories that seemed incompatible with the Biblical account of creation. Specifically, they denied the Platonic notion that the universe was created out of eternal primordial matter, a notion that compromises the sovereignty of God. The newly formulated theory of divine creation ex nihilo provided a defense for the belief in one free and transcendent Creator who is not dependent on anything. This became the rationale for asserting God's supernatural existence outside of creation and his miraculous powers that worked within the natural world. Here was a theory that had to be renounced before scientific materialism could be fully developed.

Another principal element of scientific materialism that stems from early Greek and Christian thought is the privileged role of mathematics in nature and scientific inquiry. The notion that phenomena themselves and the laws governing them are essentially quantifiable can be traced back in the Christian tradition to the writings of Augustine, who in turn drew on Plato. Augustine likened the truth and immutability of the rules of numbers to the truth and immutability of the rules of virtue; and the rules of virtue are not to be separated from the rules of wisdom, which are also true and immutable. The source and dwelling of numbers and of wisdom, he claimed, is far beyond the physical realm. God gave number to all things, and he powerfully reaches from one end of existence to the other by the power of numbers; so numbers are a part of divine wisdom. All external things have forms because they have number; they have number as their source; and they exist only insofar as they have number. Numbers exist beyond space and time, they transcend the human mind, they are as changeless as truth itself; and the wise man who beholds number and wisdom in truth itself values even himself to be of less worth than that truth.[1] The belief that God created the universe by way of the divine, transcendent language of mathematics is one that captivated the minds of many natural philosophers a millennium later.

Birthing Pains

Leaping a millennium at a single bound, we now turn to the end of the medieval period to examine some of the proximate theological influences on the rise of science and scientific materialism. During this era, the Devil figures prominently in the writings of Christian contemplatives and theologians, who appeared to be in constant fear of Satanic intrusion in prayer and all other aspects of the spiritual life. Preternatural effects were asserted by the Church to emanate ultimately from only two possible sources: God or the Devil. Benign supernatural effects, they maintained, could confidently be expected when faithful men followed the rituals prescribed by God and the Church, and they were to be found in the lives of the saints.

The Church had its own repertoire of methods for calling forth such effects, including the Mass, the healing power of saints and relics, and exorcism of the possessed. Just as the display of miracles, supernatural cures, and prophetic ability were important means of conversion in the Hebrew Bible and the New Testament, so was the claim to supernatural power an essential element in the medieval Church's fight against paganism. Working miracles and prophesying were important means for demonstrating the veracity of Roman Catholic doctrine.

Foremost among the Church's supernatural rituals were the sacraments, which were believed to work regardless of the moral worth of the officiating priest. After the sacraments came the prayers of the faithful for divine intercession. According to Christian belief, a prayer had no certainty of success and would not be granted if God chose not to concede it. A magical spell, on the other hand, was believed to work automatically if it was performed correctly. Thus, the distinction is one of supplication as opposed to a mechanical means of manipulation, or coercion. The term *magic* here refers to a system of practices in which the imagination, verbal invocations, and other rituals are performed as a means of manipulating occult forces and preternatural beings.[2] While the magical claims made for Christianity were refuted to varying degrees by the Church leaders, at the popular level they were widely embraced.

Although the practice of some types of magic was tolerated by the Church, other forms of magic were strictly banned. Indeed, those who sought to achieve marvelous results by means that were neither purely natural nor commanded by God were thought to have allied themselves, either tacitly or expressly, with Satan. The clergy thought themselves to be especially well equipped in detecting the hand of the Devil because of their training in Christian theology. This medieval attitude toward magic was clearly expressed in 1486 by the two priests Henry Kramer and James Sprenger in their influential work *Malleus Maleficarum (The Hammer of the Witches)*. Drawing from God's commandment to Moses, "Do not allow a sorceress to live,"[3] they declared it heretical to doubt the existence of witches. Although only God can perform true miracles, Satan, they maintained, has knowledge of the whole of nature and is able to perform acts that *appear* miraculous by causing effects that seem to be supernatural. On this basis they concluded that witches achieve their effects only with the aid of Satan and demons and are therefore worthy of being put to death.

Belief in the power of magic was prevalent throughout sixteenth-century Europe, based not only on biblical authority but on experience as well. In his book *Demonolatry*, published in 1595, Judge Lorraine Remy claimed that stories of witchcraft are true, *for they derive from the independent and concordant testimony of many witnesses*. The empirical facts alone, he asserted, make it "easy to understand and be fully convinced that there are witches, unless we deliberately intend to see and understand nothing."[4] While nearly all societies have believed in witchcraft, Christianity understood this in terms of the Devil and his malevolent powers. In the view of a number of prominent intellectuals of early modern Europe, violent and reckless events, such as hailstorms and other calamities, were seen as alien to God's creation—which God had deemed "good"—and were therefore attributed to diabolical influence.

During the sixteenth century conflicts arose between an emergent mechanical view of the universe and resurgent interest in more organic views, argued by the Paracelsians and others. Proponents of the latter philosophies believed that matter has the power of self-motion and of perception and "miraculous" events could occur without supernatural intervention. Moreover, action at a distance—which included the reading of minds, healing through prayer, and moving physical objects by thought alone—was seen as a natural phenomenon. According to this view, the Divine is more an *anima mundi* than an external, supernatural Creator.[5]

Both philosophies advocated "experimental philosophy," emphasizing experience, observation, and experimentation, but the mechanical philosophy also emphasized the unaided power of reason. Advocates of organic philosophy, in contrast, claimed that it is impossible for philosophers to discover the occult properties of things by use of reason; they were discoverable only through *experience*, and the reasons why things possessed those properties would always remain an unsolved mystery. The purpose of organic philosophy and its empirical methodologies, therefore, was simply to identify and use those properties; it was *not* to articulate an intelligible explanation for their underlying mechanisms.

A crucial method of Renaissance organic philosophy was the human imagination (*vis imaginativa*), which, according to the highly popular works of Paracelsus (1493–1541), was a mighty force for either good or evil. This point was corroborated by Cornelius Agrippa von Nettesheim (1487–1535), who wrote that the human mind, when "strongly elevated, and enflamed with a strong imagination," is able to cause "health, or sickness, not only in its proper body, but also in other bodies."[6]

The cosmos, according to this organic view, was pervaded by a world soul, populated by angels and demons, and subject to divine retribution and redemption, Satanic temptation, and occult influences. Its adherents promoted angelic magic, while acknowledging the power of demonic magic, and they felt this position did not pose a threat to Christianity. Moreover, for the followers of this philosophy, the summoning of celestial beings was a religious rite, in which prayer played an essential part and where piety and purity of life were deemed essential. This level of practice became a holy quest entailing a search for knowledge, not by impersonal, objective research but by individual, experiential revelation.

For scientific materialism to emerge from the womb of medieval theology and appear as a dominant force in European society, these beliefs and experiences acknowledged by the Roman Catholic Church and Renaissance organic philosophy had to be refuted. Help in this regard came from the early Protestant reformers, who claimed that the medieval Church had tried to counter popular magic by providing a rival system of ecclesiastical magic to take its place. The Protestants, in contrast, tended to disparage the whole notion of magic, as part of their onslaught on what they perceived as relics of paganism in the teachings and practice of the Roman Church. One of the early Protestant critics of magic and witch-hunting was the physician Johann Weyer (1515?-88). While acknowledging that Satan has the power to transport bodies—as he did when he tempted Jesus—Weyer argued that the witches' sabbath and their flights to these nocturnal gatherings were diabolical illusions. The arguments presented in his work *Tricks of Demons* were weakened, however, by his admission that he himself, in broad daylight, before an audience, had witnessed the levitation of a witch into the air.

Reginald Scot, a Kentish country squire, went considerably further than Weyer in his book *The Discoverie of Witchcraft* in attacking the practice of witchcraft. Writing in the late sixteenth century, he claimed that since Christ's resurrection, God had produced no more miracles and would produce no more in the future, for the Christian religion had been sufficiently established. Scot acknowledged that spirits do exist, but they are neither corporeal nor visible: an example of an evil spirit is the "spirit" of hatred, and devils are to be understood as vices. For the opponents of witchcraft it was crucial to maintain that the Devil had no temporal power, that he could not assume bodily form, and that his assaults were purely spiritual. This was the type of theological shift that was crucial for the rise of scientific materialism.

Sorcery cases in the sixteenth century often included a Catholic priest, whereas the Reformation denied the priest his magical functions and rejected the powers of supernatural intervention on the part of saints. Thus, while Catholics had saints to call on, the Protestants had only the "cunning man," or sorcerer, to look to for solutions to mundane problems. Many of the lay public believed the sorcerer was taught by God, helped by angels, or even possessed some divinity of his own. For this reason, the leaders of the Counter-Reformation sometimes explicitly associated sorcery with the rise of Protestantism. To substantiate this claim, they could point to the fact that concern with witchcraft as a real threat was a late fifteenth- and sixteenth-century phenomenon and not a major part of medieval culture. Thus, fears about witchcraft, ironically, were often greatest in the context of Protestantism.

Two attitudes were critical to the decline of belief in magic in seventeenth-century Europe. The first was the theological assertion of an orderly, regular universe unlikely to be upset by the capricious intervention of God or the Devil, which established the theological justification for the principle of universalism. This view had long been developed by theologians who emphasized the orderly way in which God governed the world, working through natural causes accessible to human investigation. Nature came to be seen as a fundamentally reasonable domain, and talk of miracles came to appear increasingly implausible. Here was another of the indispensable principles necessary for the rise of scientific materialism.

A second attitude necessary for the development of scientific materialism was the optimistic conviction that the natural causes of apparently mysterious events would one day be revealed. This faith in the power of human inquiry was bolstered by the dramatic progress made by seventeenth-century scientists and was indicative of the educated class's immense confidence in the potentiality of future human achievement. This optimism may be due in part to the relative social stability in western Europe in the late seventeenth century, in contrast to the enormous social and political upheavals of the age of witch-hunting, from the late fifteenth century through the mid–seventeenth century. The ruling class thus experienced a rising sense of confidence and a concomitant dismissal of preternatural interference in the course of nature and society. Scientists of this time exuded great confidence that they were entering a new age in which all of nature's secrets would be revealed and thereby offer to humanity total power over the natural world. Thus, the emerging mechanical philosophy triumphed over natural magic in part because it was regarded as an "establishment" philosophy that upheld religion and the social order. Further, it not only legitimated but showed the feasibility of the mechanical appropriation of the natural world without impugning the miraculous nature of Christ's works.

Before moving on to the early development of the mechanical philosophy of modern science, one further influence of the Reformation on the Scientific Revolution merits attention. Many of the Protestant reformers taught that the way to solve practical difficulties is through a combination of self-help and supplication to God. Magic, including that endorsed by Roman Catholicism and the organic philosophy of the day, was condemned as both impious and useless. Mysterious uses of the imagination as a means of dealing with problems were spurned not only because they made one prone to diabolical influence but because they were thought to be *too easy*. According to the new Protestant ethic, practical problems were to be solved through honest, hard, physical work—a view that could be seen as a theological move toward physicalism. This attitude encouraged people to seek technological solutions to their mundane problems rather than magical ones. In short, the deeply rooted interests of many thinkers in the seventeenth century called for the systematic, rational, and empirical study of nature for the glorification of God in his works and the control of the corrupt world.[7] Western Christian theology in general and the Protestant Reformation in particular thus played a major role in establishing the theological grounds for the Scientific Revolution and the eventual ascendancy of scientific materialism.

The Infancy of Scientific Materialism

The mechanical philosophy of René Descartes (1596–1650) provided a fitting match to many of the theological trends of the seventeenth century. Articulating the widespread inclination to disenchant the universe, which may have been driven by a desperate need to end the cultural hysteria of the era of witch-hunting, Descartes asserted that "there exists nothing in the whole of nature which cannot be explained in terms of purely corporeal causes, totally devoid of mind and thought."[8] Descartes introduced two major exceptions to this principle: (1) biblical miracles have no mechanical explanations and (2) the human mind, which he equated with the soul, is an immaterial, immortal gift from God. In this way the closure principle was introduced into Western modernity, but with the two exceptions of miraculous divine intervention in nature and the influences of the human soul.

In viewing the world as mechanism, devoid of psychic contents such as the sense of heat and pain, Descartes rejected the ontology of the Scholastic tradition. For the sake of clarity and distinctness, he deemed it necessary to step outside ourselves and take a disengaged perspective—that is, to seek an objective, God's-eye view. This philosophical move had a scientific corollary in Copernicus's decision to view the motion of the earth not from a human, terrestrial perspective, as was the habit of traditional astronomy, but from the objective, Godlike perspective of the sun. The attempt in both cases was to view reality from a vantage point that transcended the limitations of human subjectivity. Thus, Descartes was one of the first modern champions of the ideal of objectivism, perhaps the most central of the principles of scientific materialism.

The cosmos as conceived by Descartes and his followers was entirely explicable in terms of inert matter, and it was the task of the natural philosopher to articulate the underlying processes of the natural world. In this view nature was seen as a domain in which neither angels nor demons nor even God interfered, though a crucial theme of Descartes's philosophy was his insistence that bodies continue in existence only because God preserves them in being. Most nonmechanical events were confined to the distant past—either to biblical events or to occurrences at Creation that were omitted in the biblical narrative. The sole current exception to this rule was the divine infusion of a human soul into the body, which continued to influence the body through the pineal gland. Descartes theorized that the pineal gland, on decision of the soul, creates voluntary actions of the body in a purely mechanical way, while all other actions are reflexive. By confining the physical causal efficacy of the soul to its interaction with the pineal gland, the rest of the body, including the nervous system, muscles, and so on, could be regarded purely as a mechanism, quite devoid of any non-physical elements. Thus, the physiological imagination could be applied to all aspects of the functioning body except the pineal gland. The influence of this presumed special status of the pineal gland seems to have lingered well into modern times. Until the last three decades of the twentieth century, the pineal gland was uniquely neglected by physiological and biochemical investigators; whereas for more than three hundred years the rest of the nervous system and body has been "fair game" for physiological research. Nevertheless, as recently as the late nineteenth century, many experts agreed with Aristotle's view that the function of the brain is to cool the body.

Leibniz (1646–1716) removed this single, persistent anomaly of the soul's "special relation" with the pineal gland by declaring that the human soul does not act on the body; rather, God causes the body's mechanical movements to conform to the will of the soul according to a pre-established harmony, an original miracle, that was wrought at Creation. Mind and spirit were now denied all causal efficacy in nature, and the closure principle was presented as a universal law of nature. Far from tending toward atheism, the mechanical philosophers believed that their assertion of the mind and life originating outside of nature and exerting no influence upon it actually pointed to the existence and creative power of God, the Creator of nature. Now physical processes alone were regarded as having causal efficacy in the natural world.

This mechanical philosophy opened the way for people to feel justified in exploring, understanding, and controlling nature, without fear of diabolical association, which was crucial, given the trauma of the witch-hunting craze that dominated Europe for almost two centuries immediately prior to the rise of the mechanical view of the universe. These themes were further emphasized by Francis Bacon (1561–1626), whose experimental philosophy asserted that the human quest for power over nature was divinely sanctioned and was to be accomplished not with the use of *vis imaginativa* but with hard, physical work. "Toward the effecting of works," he declared, "all that man can do is put together or put asunder natural bodies. The rest is done by nature working within."[9] Moreover, Bacon asserted that the rise of experimental science was sanctioned by God himself, and he presented biblical evidence in support of this view.[10] The gist of his argument was that science would help restore the dominion of man over nature that was lost through Adam's sin. Along the same lines, Descartes predicted that by knowing the forces and the actions of material bodies, we can "make ourselves the masters and possessors of nature."[11] Following this methodology, the experimental philosopher could fill the shoes of the natural magician free of fear of accusation of diabolical sorcery.

Scientific Materialism's Defense of Christian Theology

Christian theology, especially its doctrine of creation, is the primary source of the non-Greek elements that informed the views of the post-Reformation natural philosophers.[12] While belief in the immanent laws of nature is characteristic of classical Greek views, belief in divinely imposed laws is best illustrated in Jewish monotheism. By the seventeenth century, the ideals and values of the precursor of scientific materialism were well integrated with and supported by the ideals and values of especially the Protestant Church. Rationalism and empiricism, for example, were central to both the Puritan ethic and to science; and leading figures of the nuclear group of the Royal Society were deeply religious men, markedly influenced by Puritan conceptions. Similarly, Protestant academies in France attended more to science than did Roman Catholic ones, and Pietistic secondary schools and universities in Germany characteristically leaned toward science. The most eminent philosophers of nature during the seventeenth century were not members of the clergy but laymen who developed their own theologies. Rather than uniformly adhering to established Roman Catholic or Protestant theologies, they sifted through the diversity of Christian theories of their times and emphasized views of their own choosing. In hindsight they appear to be pioneers of their own sect of Christianity, which was to evolve into the ideology now known as scientific materialism.

The Protestant reformers and the founders of the new mechanical philosophy joined forces to banish the Renaissance organic, animistic view of nature. They recognized that such animistic beliefs concerning matter could imply that the death of the human body implies the disappearance of the soul. The mechanical philosophy, in contrast, asserted the supernatural infusion of the soul at conception and its extraction at death and was thus seen as offering a defense against that heresy. Moreover, the animistic notion of self-moving matter could imply a self-organizing universe that might account for the order of the world without resort to an external Creator. Once again the mechanical philosophy seemingly came to the rescue.

Isaac Newton was the most prominent champion of the view that the universe, understood as composed of inert bits of matter, required an external God who created matter, put it in motion, and imposed laws upon it. Robert Boyle, too, strongly advocated the biblical assertion that humans are made in the image of God, not nature, and this undermined the organic model of nature, which drew analogies between microcosm and macrocosm and between humans and the rest of creation. The new experimental philosophy also countered the Paracelsians' emphasis on inner illumination, which could be perceived as a threat to established forms of religion and could also threaten the collaborative aspects of scientific research. Thus, a number of the advocates of the mechanical view of the universe believed that to defend Protestant Christian doctrine and to work for the good of the public, objective research into the external universe beyond the veil of appearances was to be promoted; while first-person, phenomenological modes of inquiry into the nature of appearances themselves, as advocated in the organic view of nature, were to be suppressed.

Finally, the animistic belief in action at a distance and in the occurrence of miraculous events devoid of preternatural intervention undermined the "argument from miracles," which was a central pillar of both the Roman Catholic and the Protestant Church's authority. The mechanical philosophy once again pointed the way to a solution: all physical influence must be due either to contact by another physical body or to supernatural intervention. It was in this metaphysical context that Newton was so concerned about the apparent action at a distance involved in gravitational attraction. His theological response was that God imposed this appearance of mutual attraction on matter, in accord with Newton's more general belief that God imposed all natural laws on the material world. Thus Christian theology and the mechanical philosophy agreed that action at a distance could not be both natural and miraculous.

More generally, Christian theology found it imperative to draw a strict distinction between true miracles, those resulting from divine, supernatural intervention, and mere marvels or extraordinary events. Miracles presupposed a natural order against which they could be judged miracles and thereby distinguished from mere marvels. In the seventeenth century, Christian theology and the mechanical philosophy of nature took the common position that an event could be deemed a miracle if it was not explicable in terms of natural laws. Science was designated as *the* proper mode of inquiry to determine those laws, which made the identification of miracles possible. Thus, the mechanical philosophy of the seventeenth century—which may be regarded as the infant stage of contemporary scientific materialism—was seen as a crucial support for Christian theology, and many of its principles were largely theological in origin.

The Judeo-Christian concept of imposed laws of nature, which burst into prominence in seventeenth-century scientific thought, can be traced to the latter part of the thirteenth century, when a new tradition of Christian theology arose, called the theory of voluntarist natural law. This theory conceived of law as imposed on the world by the divine will. This was strongly set forth in the ethical voluntarism of William of Ockham, who grounded the natural law of morality on the will of God. Natural law, therefore, became a divine command, which is right and binding merely because God is the lawgiver.[13] According to Ockham, there are no necessary intermediaries between an infinitely free and omnipotent God and the things that he has created and that are utterly contingent upon him. So the order of the world can be discovered only by an examination of phenomena, not by any a priori reasoning other than the careful examination of God's revealed word. Given his *absolute* power, God could order the opposites of the acts that he has in fact forbidden. However, by his *ordained* power, he has actually established a moral order, within the framework of which the natural law is absolute and immutable. Thus, God was thought to have made a pact, or covenant, with his creatures to abide by the moral laws he had imposed on creation.

The theory of imposed natural law seems to have made its way into Protestant thought through Martin Luther (1483–1546), who was aware of medieval thinking on this subject, and Ulrich Zwingli (1484–1531) and John Calvin (1509–1564) also embraced this notion. Puritan theology, for example, drew a distinction between the ordinary and extraordinary Providences of God that is strongly reminiscent of the voluntarist theories of Ockham. Descartes, Leibniz, and Locke, as well as Boyle, Newton, and other members of the Royal Society, proceeded to adapt the voluntarist theory of moral law to a comparable view of physical laws operating in nature. Many of these later philosophers and scientists were well aware of the writings of the medieval voluntarist theologians. Newton believed that the divinely imposed laws of nature changed from place to place and from time to time throughout the universe, as if God were experimenting with his creation in different ways at different times and places. However, shortly after Newton's time, an increasing number of natural philosophers chose instead to believe that God imposed his laws on nature in a manner that was the same throughout space and time. In making this move, they established the principle of universalism.

The new mechanical philosophy asserted that individual entities are the ultimate constituents of nature and possess no inherent connections with one another. Each exists in total isolation from the rest, the relations between them are imposed on them from without, and these imposed patterns of behavior are the laws of nature. In this way the principle of physical reductionism, yet another of the central pillars of scientific materialism rooted in Christian theology, was established. According to Robert Boyle, the laws of motion did not arise from the nature of matter but were imposed on the world by the will of the Creator. The ordinary course of things can be abrogated, as in the case of miracles, by the Creator alone (or agents assisted with his absolute or supernatural power); for God, being omnipotent, can do whatever involves no contradiction. Boyle, like Newton, was very concerned to demonstrate that the universe was not to be seen as the self-sufficient, active, and productive source of all things. Rather, the laws according to which nature operates are ones that God freely chose, and his divine activity retains a crucial role in the world.

A number of Protestant reformers gave strong theological sanction to the scientific study of nature. Calvin, for example, declared that God's radiant glory pervades and upholds the universe to such an extent that one could even say that "nature is God."[14] Many natural philosophers of the late seventeenth century, including Boyle, shared a corresponding vision of scientific inquiry itself as a form of worship, with scientists serving as priests in the temple of Nature. Similarly, Newton regarded God as an omnipotent cosmic sovereign who governs all things as the Lord over all. In the very laws he discovered, Newton saw a proof of God's continued presence in the world; with enormous self-assurance, Newton hoped to eliminate disputation both in natural philosophy and in biblical exegesis by achieving definitive truth. Thus, these pioneering natural philosophers looked to science to be the final arbiter to help settle theological controversies.

The Christian Theological Defense of Scientific Materialism

Despite the extensive common ground between Christian theology and the mechanical philosophy that was emerging in the seventeenth century, many natural philosophers and theologians of this time saw a potential threat to Christianity from this new way of viewing nature. Thus, while some theologians looked to the mechanical philosophy to defend Christian doctrine against the potential heresies of organic views of nature, other theologians and natural philosophers believed that this mechanical view needed to be defended against those of its critics who thought it might lead to heresy.

A revealing historic example of this impulse is found in Thomas Sprat's defense of the mechanical and experimental philosophy against those who suspected it of undermining the one true religion. In 1663, Sprat was nominated for membership in the Royal Society with the provision that he would write a history of the Society to help it defend itself against its religious detractors. In his work *The History of the Royal Society of London* (1667) he openly declared that the principal purpose of mechanical/experimental philosophy was to acquire power over nature and that this endeavor was focused on matter. Furthermore, he argued that, of all pursuits, such scientific research was most likely to engender a spirit of piety, perseverance, and humility—the hallmarks of Christian virtue. Thus, science could offer sanctuary from theological disputes as well as present its own brand of sanctification. In his view, the Protestant Reformation and the new philosophy of nature had this in common: each prized the original copies of God's two books, nature and the Bible, while being intent on bypassing the corrupting, subjective influence of scholars and priests.

Far from denying the existence of invisible beings, Sprat asserted, this material pursuit reinforced the philosopher's belief in invisible beings because the infinite subtlety of parts of matter cannot be detected by even the sharpest senses. With this assertion Sprat, expressing a view common among scientists of his day, established the methodological guidelines for replacing the preternatural realm of demons, spirits, and angels with the modern scientific theoretical realm of quarks, virtual particles, and superstrings. The old domain of theoretical entities was gradually repopulated by the new.

Proponents of the new philosophy were presented with a dilemma: to support the experimental methodology of science, many early natural philosophers affirmed the voluntarist view of natural laws being imposed on nature from without. Yet if God were to intervene miraculously in the course of nature whenever, and in whatever manner, he freely chooses, then there would be no metaphysical grounds for empirically determining universally consistent laws of nature. Another related problem remained: How could scientists fulfill their central aim of controlling nature if it was still under the voluntary control of God? Sprat, in accord with much of the theological and philosophical thinking of his day, resolved both problems by declaring that God seldom or never chooses to perform miracles in times when natural knowledge prevails but does offer miracles to dark and ignorant ages. With this ingenious theological ploy, he assured his readers that God would not violate the closure principle that was to become one of the central articles of faith of scientific materialism. Sprat maintained that the experimental philosopher has no need of miracles, for he sees impressions of God in *all* of nature. We can only speculate as to whether Sprat was aware of the ease with which he moved from the statement that miracles exist everywhere to the assertion that miracles exist nowhere. In any case, the omnipotent Creator of the universe was now seen as having freely chosen not to intervene any longer in the world, thereby making himself, in effect, impotent.

With respect to theological claims of God's decrees, his immateriality, and eternity, Sprat claimed that experimental philosophers are satisfied with a plain believing, or unquestioning faith, requiring no empirical evidence or experiential confirmation. Thus, he argued, Christian beliefs should be regarded as safe and even strengthened in the hands of natural philosophers. Once religion is identified with uncritical belief based on revelation, and science with knowledge drawn from active, human inquiry, the stage is set for science to begin challenging all religious doctrines pertaining to nature.

With respect to the Devil, Sprat, again assuming an authority that is normally granted only to supernatural revelation, assured his readers that Christianity is secure, so the Devil is no longer a threat. All other spiritual entities, such as spirits, demons, and fairies, are illusions, and their nonexistence in nature has been demonstrated, he claimed, by *experiments*, though he did not specify exactly which experiments he had in mind.[15] Newton, on the other hand, did not do away with such ghostly entities altogether. Rather, he applied a systematic interpretation of biblical references to such preternatural beings; cherubim and seraphim were hieroglyphs of ordinary social groups, evil spirits were mental disorders, and devils were imaginary ghosts of the departed. Thus, mysterious spiritual entities, which had been previously thought of (and occasionally perceived) as roaming in the objective world of nature, were now quarantined in the subjective world of human society and consciousness. God's outer creation had now been cleansed of these contaminating influences, leaving only the inner being of man defiled. It would take another two hundred years before Western psychoanalysts would have the nerve to begin the scientific exploration of these dark inner realities. Two quite different reasons seemed to contribute to the human soul being removed from the domain of science: it was beyond the scope of science, for it was an immaterial, immortal gift infused into man by God; and it was corrupt, for it was sinful to its core, wherein lurked all manner of evil spirits in the forms of neuroses and psychoses, which the modern psychoanalytic tradition was eventually to classify and seek to explain.

With his theological defense of the mechanical philosophy, Sprat appears to be one of the pioneering theologians of this new sect. Without introducing any rational or empirical evidence, he dispensed with the preternatural realm altogether, declared that God was ineffectual in nature, and decreed that religious convictions, unlike scientific assertions, were solely matters of belief. With these new metaphysical principles, the mechanical philosophy was now prepared to focus on the material world exclusively and to leave its host society with the impression that this was the only area in which genuine, hard-won knowledge was to be acquired.

The Triumph of Scientific Materialism

The original quest of scientific materialism was to seek a God's-eye view of the universe, and because God was regarded as utterly transcending man, the world as seen by God must transcend the world of human experience. The natural philosophers of this era, who were in many instances avid theologians, envisioned knowing the mind of God through knowing his Creation. The culmination of this scientific quest might even be seen as a kind of apotheosis, when man's understanding of the natural world merged with the understanding of God. The same goal, which had been promised by Roman Catholic theology as the culmination of the contemplative life *in heaven*, was now seemingly brought within reach *on earth* due to the methods of the new experimental philosophy. This goal was regarded as all the more plausible as these natural philosophers conceived of God in their own image, as divinely skilled in mechanics and geometry.

While many scientists of the seventeenth century were deeply concerned with affirming God's continuing role, as not only the Creator but also the active Governor of nature, the subsequent development of science seemed increasingly to undermine that theological supposition. Miracles were widely thought to be confined to the distant past, and in relation to the mechanical workings of nature, the divine came to be seen as a "God-of-the-gaps." As the gaps in scientific understanding of the mechanics of nature closed, there seemed to be fewer and fewer openings for God's supernatural intervention. And, as scientific knowledge progressed, more and more people felt such confidence in this mode of inquiry that they assumed the remaining gaps would eventually be filled by rational and empirical understanding without resorting to the assertion of divine intervention. Thus, already at the beginning of the nineteenth century, when asked about God's role in governing the heavens, the French astronomer Pierre Laplace is reputed to have replied, "I have no need for that hypothesis."

In retrospect, it seems almost inevitable that Descartes's assertion that the mind exists independently of the body was bound to be challenged by science, thereby calling into question the very existence of the mind as a real entity. Already at the close of the seventeenth century, theories of material souls were being formulated to replace earlier notions of a spiritual soul. And the mechanization of the mind became a familiar theme in the clandestine literature of the Enlightenment. Indeed, as Newton strenuously argued, Leibniz's view of a self-contained, self-sufficient universe, free of all spiritual influence, was bound to lead to materialism and atheism.[16]

Since the early eighteenth century, Christian theology has been on the defensive against the onslaught of scientific knowledge. For many people who have embraced scientific materialism, traditional forms of religion have been reduced to socially acceptable formulas with which to embellish a life that has been made comfortable by science and technology. Because of the pervasive influence of scientific materialism, many Christians today have come to regard most of the creed of Christianity as it existed prior to the Scientific Revolution as unnecessary, even as a mere hindrance to the religious life. Contrary to Thomas Sprat's assertion that Christianity was utterly secure and that its beliefs were safe, and even strengthened, under the care of the proponents of the new mechanical philosophy, history has proven this true only insofar as one identifies Christianity with the specific theological creed of scientific materialism.

In the development of Protestant Christianity following the Scientific Revolution, it appears that the movements that have most enthusiastically emphasized personal religious experience, such as the Pentecostals and Quakers, also represent a mentality that is furthest removed from the methods, ideals, and worldview of science. By the nineteenth century, religion in the West had come to be strongly associated with Romanticism: it dealt with matters of the heart, leaving matters of the intellect to science and philosophy. Thus, religion, which exalted faith and tradition, often presented itself as independent of reason and incommensurable with science. And science, which considered itself to be based entirely on consensual, objective experience, gradually came to ignore religion on the grounds that it is intrinsically private, subjective, and even irrational in nature. While scientific materialism, inspired by the mechanical philosophy formulated during the Scientific Revolution, has triumphed in the modern West, various "organic" philosophies continue to manifest in a number of religious, medical, philosophical, and literary currents in Western culture.

In light of the symbiotic relation between the Scientific Reformation and the Protestant Reformation, modern scientific materialism may be regarded as the disenchanted heir of early Protestantism. The accommodation, demarcation, and alliance of Protestant theology and modern science lasted one and a half centuries, breaking down only after Darwinian theory refuted the idea of a static world being governed by certain and irrevocable laws. However, the secularization, or disenchantment, of the world was already set in motion in the writings of Calvin, Descartes, and Leibniz.[17]

This account of the interaction between Christian theology and the rise of scientific materialism certainly overlooks many important elements of the immensely complex influences contributing to the Protestant Reformation and the Scientific Revolution. It does not address, for instance, the influences of Jewish and Arabic thought or the significance of the Roman Catholic Counter-Reformation in the rise of modern science. Nor does it comment on the multiplicities of views among proponents of diverse organic and mechanical philosophies. Leibniz, for instance, who promoted the closure principle, was a major advocate of an organic view of nature. Newton was fascinated by mysticism, advocated a number of principles of an organic philosophy, and devoted more of the closing two decades of his life to writing theology than to scientific research. But such incongruous facts as Kepler's work in astrology and Newton's research in alchemy are commonly sanitized away in histories of science in order for these figures to be presented as modern heroes of rationality who illuminated the way out of the superstitious magic and religion of the Dark Ages.

The crucial point of this historical sketch—which I hope is not undermined by its brevity and simplicity—is that the principles of contemporary scientific materialism were laid down long before there was any compelling scientific evidence to support them and that the justifications for adopting these principles were largely theological in nature. And it is the same principles—commonly presented as purely objective, scientific truths—that dominated most scientific empirical and theoretical research in the twentieth century. Indeed, their advocates adhere to them with all the tenacity of religious believers everywhere. One may argue for the validity of these principles on the grounds that research conducted within their conceptual framework has been extremely successful in terms of understanding the objective world and providing humanity with a wealth of technology. This is certainly true. But it is equally true that scientific materialism has obscured the subjective world pertaining to the nature, origins, and function of the mind and consciousness and has obscured the relation between the inner world of consciousness and the outer world of the objects of consciousness. In order to expand the scope of scientific research into the domain of consciousness and other subjective realities, science must be released from the metaphysical shackles of scientific materialism. This will call for a noetic revolution every bit as radical and profound as that which opened the way for the rise of modern science itself.

PART II

Toward a Noetic Revolution

AN EMPIRICAL
ALTERNATIVE

A Return to Experience

Occam's Razor: It is vain to do with more assumptions what can be done with fewer assumptions.

Scientific materialism is riddled with assumptions stemming from the absolute dualisms of Descartes's mechanical philosophy: the dualism between sensory appearances and physical reality, between subject and object, and between mind and matter. On the whole, scientists since his day have focused their research on the domain he designated for them: the objective material world. These are the topics that have interested them and captured their attention, while sensory and mental experience and consciousness itself have been largely ignored. Thus, in accordance with the relation between attention and experienced reality suggested by William James, only objective, physical phenomena and their attributes have come to be regarded as real; while subjective, mental phenomena have come to be treated as "mere waste, equivalent to nothing at all."[1]

Scientific materialism has served admirably as a metaphysical framework for the scientific investigation of external, physical phenomena, but it has proven inadequate as a framework for the scientific investigation of internal, mental phenomena. For such research we must explore what can be done with fewer assumptions, namely, fewer assumptions than those of scientific materialism, which have been inhibiting the scientific study of subjective reality for centuries. For this I draw inspiration from William James, who stands as a modern pioneer of the scientific study of the mind. Trained in chemistry, biology, and medicine, he offered at Harvard the first course on physiological psychology to be presented in the United States; he also founded the first laboratory for experimental psychology in this country. But such experimental psychology failed to sustain his interest, for he found that the more "progress" there was with this mechanistic approach, the more disappointing and trivial were its conclusions. James was a premier example of a man of science who refused to adhere to the articles of faith of scientific materialism and a deeply religious man who rejected all religious dogma. His approach was to take a genuinely scientific interest in the precise, open-minded investigation of the entire range of human experience, including religious experience.[2]

The adoption of the kind of empiricism envisioned by James demands that we regard even our most assured conclusions concerning matters of fact as hypotheses liable to modification in the course of future experience. He calls his perspective radical, for it does not admit into its construction any element that is not directly experienced, nor does it exclude any element that is directly experienced. Thus, ordinary, commonsense, firsthand experience is taken as seriously as scientific, third-person observations, and even the most cherished principles of scientific materialism, such as monism, are treated simply as hypotheses.[3] James also rejects the principle of reductionism with his assertion that *"the relations that connect experiences must themselves be experienced relations, and any kind of relation experienced must be accounted as 'real' as anything else in the system."*[4] In his view, the directly apprehended universe needs no extraneous trans-empirical connective support but possesses its own coherent, continuous structure. Moreover, he gives no credence to the existence of any absolute mental or physical substratum to the world of experience.

By means of conceptual analysis we may consider any number of hypotheses, but James believed the terminus of thought must be perception, by which we retroactively validate earlier virtual knowledge that was not drawn from experience. Most of the time, however, our conceptual understanding goes unchallenged and thereby substitutes for knowing in the most complete sense. Moreover, experience itself must be carefully monitored to determine what is actually being perceived and what is being conceptually superimposed upon experience. As noted earlier, in the sixteenth century Johann Weyer declared that he had witnessed the levitation of a witch into the air; and in the seventeenth century the Dutch naturalist Anton van Leeuwenhoek and his male scientific colleagues reported seeing fully formed little humans swimming around when they examined semen under their new microscope. More recently psychologist John Anderson has asserted evidence of computer systems displaying frustration,[5] and journalist Michael Lemonick has described a computer-generated image of the brain based on a positron-emission tomography scan as a sad thought.[6] All these people were certainly observing *something*, but what they were *thought they were seeing* was determined largely by their assumptions and expectations.[7]

One of the most salient instances in the history of science of the maxim "believing is seeing" occurred at the beginning of the twentieth century. In

1903, René Prosper Blondlot (1849–1930), a distinguished French physicist, claimed to have discovered a new type of radiation while trying to polarize X-rays, which had recently been discovered by Wilhelm Roentgen. Blondlot called this new radiation the N-ray, after Nancy, the name of the town and the university where he lived and worked. He claimed he had generated N-rays using a hot wire inside an iron tube. These rays were detected by a calcium sulfide thread that glowed slightly in the dark when the rays were refracted through a 60 degree angle prism of aluminum. A narrow stream of N-rays was thought to be refracted through the prism and produced a spectrum on a field. Blondlot reported that the N-rays were invisible, except when viewed as they hit the treated thread. When he moved the thread across the gap where the N-rays came through, the thread was illuminated, and he attributed this to N-rays. Based on his empirical findings, Blondlot concluded that N-rays are emitted by all substances except green wood and certain treated metals. Following the report of this remarkable discovery, dozens of other scientists followed his experimental procedures and confirmed the existence of this new type of radiation in their own laboratories.

Laboratories in England and Germany, however, had not been able to replicate Blondlot's results, which led *Nature* magazine to send the American physicist Robert W. Wood of Johns Hopkins University to investigate Blondlot's findings. Wood was skeptical of Blondlot's claims. To put his suspicions to the test, unbeknownst to Blondlot or his assistant, he removed the prism from the N-ray detection device, without which the device couldn't work. Yet, when Blondlot's assistant next ran the experiment without the prism, he found that it produced the same positive results as before! When Wood then tried to surreptitiously replace the prism, the assistant saw him and thought he was removing the prism. So the next time the assistant ran the experiment, he claimed he could not see any N-rays. But he should have, since the equipment was in full working order.

Blondlot, his assistant, and dozens of other scientists who replicated Blondlot's findings in their own laboratories are now said to have suffered from from self-induced visual hallucinations, an affliction that is never clearly defined. Exactly how they slipped into this delusion, and what made them prone to replicating each others' hallucinations has never been adequately explained. The book was quietly closed on this embarrassing episode, and on those rare occasions when it is mentioned by scientific materialists, the moral they draw from this story is that even though scientists often make errors, even big ones, other scientists will uncover the errors and get science back on the right path to understanding nature.[8] But how are we to discern errors of perception that cannot be revealed with such a simple maneuver as removing a prism? To what extent are expectations and beliefs structuring purportedly objective, scientific observations, let alone theorizing, as a whole? The only reason the fallacy of Blondlot's findings was discovered was because someone was *skeptical*. But, as I said earlier, we can't be skeptical of something of which we are not even *aware*. And if current scientific, philosophical, journalistic, and pedagogical writings on the mind/body problem are any indication, many people seem quite unaware of the degree to which researchers commonly conflate the assumptions of scientific materialism with the empirical facts of scientific research. The purpose of raising such questions is not to undermine the credibility of science but to encourage the healthy note of skepticism concerning unchallenged assumptions that has always helped sustain the rigor of scientific inquiry.

In the process of growing up, and in the process of learning to make scientific observations, we learn to see by doing, not just by looking; and our actions are directed by our beliefs. For example, it may seem at first glance that a biologist is simply observing whatever is seen through a microscope. But first one must ask what kind of a microscope is being used. Is it an ultraviolet microscope, a phase contrast microscope, a interference contrast microscope, an X-ray microscope, an electron microscope, or an acoustic microscope? Since the early twentieth century, even the conventional light microscope has essentially been a Fourier synthesizer of first- or even second-order diffractions. Thus, we must either modify our notion of seeing or hold that we never *see* through a serious microscope; for the "normal" physics of seeing is seldom used in observing living materials through such a device. In short, we do not see *through* a microscope, we see *with* one; even then, when it comes to biological microscopy, we are blind without a sufficient theoretical training in practical biochemistry.[9]

Psychologist Jerome Bruner comments in this regard that

> perception is to some unspecifiable degree an instrument of the world as we have structured it by our expectancies. Moreover, it is characteristic of complex perceptual processes that they tend where possible to assimilate whatever is seen or heard to what is expected.[10]

To what extent have the assumptions of scientific materialism not only limited scientific research but introduced distortions or even delusions into this mode of inquiry? To what extent has adherence to this dogma mistakenly led us to believe that we know things of which we are actually ignorant? In his book *The Discoverers: A History of Man's Search to Know His World and Himself*, Daniel J. Boorstin refers to "the illusions of knowledge" as the principal obstacles to discovery. The great discoverers of the past, he declares, "had to battle against the current 'facts' and dogmas of the learned."[11] If in the future there are to be great discoverers of the nature, origins, and functions of consciousness, they will need to demonstrate a high degree of skepticism regarding many of the purported facts and dogmas of the creed that presently dominates virtually all of scientific research.

A World of Human Experience

In contrast to the Cartesian distinction between the objective physical world and subjective experience, William James redirects our attention back to the immediate world of human experience. With his assertion that we observe external objects directly yet fallibly, he abandons the absolute distinction between the primary qualities of the physical world as opposed to sensory impressions, which have been excluded from nature; he also rejects the assertion of scientific realism that the objects we perceive exist independently of our perceptions. Thus, instead of discarding sensory impressions as being misleading, false, or nonexistent, he accepts them as they are—as the contents of the world of human experience.

James's philosophy of radical empiricism rejects the absolute duality of mind and matter in favor of a world of experience, in which consciousness *as an entity*, in and of itself, does not exist; nor is it a function of matter, for matter *as an entity*, in and of itself, does not exist either. According to this view, the postulation of mental and physical substances is a conceptual construct, as is the metaphysical distinction between subject and object. Mind and matter are constructs, whereas pure experience, which is neutral between the two, is primordial. One implication of the hypothesis that we are directly acquainted with reality is that the contents of consciousness can no longer be regarded as being "in the mind" (let alone in the brain). Reality just *is* the flux of experience.

James's radical empiricism is fundamentally at odds with scientific materialism, which assumed from the outset the absolute distinction between primary and secondary qualities. In the twentieth century, however, many scientific materialists came to the conclusion, on rationalistic grounds, that neither sensory nor mental experiences exist at all. Only the purely objective, physical world of science is real. But what, then, are we to make of the commonsense objects—with their colors, smells, and so on—that we perceive? On the same grounds that subjective experiences are denied existence—namely, because they cannot be reduced to the objective world of science—the objects that fill the world of our everyday experience might also be denied existence. In fact, some modern advocates of scientism take this final step of denying the existence of everything that appears to the common person.[12] Here is the ultimate triumph of dogmatic rationalism over experience, and there could hardly be any metaphysical doctrine more incompatible with science.

With the rejection of the intrinsic, independent existence of any phenomenon within the world of experience, there is no longer any place for the Cartesian distinction between (1) primary properties that things have in themselves apart from any contribution made by language or the mind and (2) secondary properties that exist only in relation to subjective experience. The rejection of the absolute dichotomy of primary and secondary properties further implies the rejection of the absolute dichotomy of subjective versus objective statements. Subjective and objective statements, together with conventional and factual statements, rather occur along a qualitative continuum. In the words of philosopher Hilary Putnam, whose own writings are inspired in part by the work of William James, "[w]hat is factual and what is conventional is a matter of degree; we cannot say, 'These and these elements of the world are the raw facts; the rest is convention, or a mixture of these raw facts with convention.' "[13]

If all valid statements concerning the world of human experience have both a conventional and a factual element, it follows that the referents of language are also inseparable fusions of convention and reality. Thus, the existence of a concrete object like a tree is also a matter of convention; and our observation of a tree is possible only in dependence on a conceptual scheme. The reason for this, according to Putnam, is that "elements of what we call 'language' or 'mind' *penetrate so deeply into what we call 'reality' that the very project of representing ourselves as being 'mappers' of something 'language-independent' is fatally compromised from the very start.*"[14]

In this view the subjective and objective poles of the continuum are vacuous. There is no way to justify the assertion that anything posited is purely objective or purely subjective. The world of human experience consists of a fusion of both elements, or better said, a primordial nonduality of those elements. Similarly, the "fact that a truth is toward the 'conventional' end of the convention-fact continuum does not mean that it is absolutely conventional—a truth by stipulation, free of every element of fact."[15] This assertion by no means implies that such dualistic notions as subject and object are useless. On the contrary, they point out a practical distinction that is of great importance; but this distinction is only functional, not ontological as understood by the traditional dualism of scientific materialism.[16]

We are now in a position to challenge the principle of objectivism on the grounds that "object" itself has many uses and meanings. Such terms as *object, existence, reference, meaning, reason, knowledge, observation*, and *experience* each has a multitude of different uses, and none has a single absolute meaning to which priority must be granted. Since these terms are not self-defining, we employ their definitions according to the conceptual schemes of our choice. That is, we choose our definitions; they are not determined by objective reality. On the other hand, while our choices are culturally relative, they are not decided by culture alone, nor are they arbitrary. Thus, the "same world" can be described by science and common sense, *without* trying to reduce them to a single "real" version posited as being true independently of our choice of concepts.

Once we have chosen a conceptual scheme, there are facts to be discovered and not legislated by our language or concepts. Our conceptual scheme restricts the range of descriptions available to us, but it does not predetermine the answers to our questions. As Putnam comments,

> the stars are indeed independent of our minds in the sense of being causally independent; we did not make the stars. . . . The fact that there is no one metaphysically privileged description of the universe does not mean that the universe depends on our minds.[17]

On the other hand, if there were no language users, there would not be anything true or anything with sense or reference. Thus, the rich and ever-growing collection of truths about the world is the product of the experienced world, with language users playing a creative role in the process of production.[18]

Putnam's example of mereological sums, or objects consisting of sums of other objects, well illustrates this point. The solar system is an obvious example of such a sum, or "discontinuous object," but on further examination it becomes apparent that almost all the objects we talk about fall into this category. Mereological sums exist not only as configurations of physical objects in space but as sequences, or continua, of mental and physical events in time. At what point does one object become a property or a component of another object? When we designate it as such, on the basis of our choice of definitions. Thus, our choice of conceptual schemes plays an essential role in determining what counts as an object and what does not. A shift in conceptual schemes therefore implies a shift in what we experience; and that is all we are in any position to know or discuss.

This point relates directly to the status of the mind in science as a result of the relation between attention and experienced reality. The conceptual schemas of scientific materialism omit the subjective mind from nature and thus deny its causal efficacy in the physical world. Thus, the attention of natural scientists is drawn away from the mind; and consciousness, being "out of sight," naturally drops out of scientifically experienced reality altogether. How this occurs philosophically is that the terms *entity* and *object* are so defined that the subjective mind is excluded, and thus its reality is denied. Those definitions are not "metaphysically privileged," for they are created and chosen by humans, not by objective reality.

Given James's assertion that we perceive external objects directly and fallibly, as opposed to inferring them on the basis of subjective appearances, our task is to make sense of the phenomena from within our world, rather than to seek a God's-eye view or a view from nowhere. Drawing on the insights of quantum mechanics and modern logic, Putnam argues that as the circle of science gets larger, paradoxes emerge that demonstrate that a God's-eye view, or a view from nowhere, is impossible in principle. In quantum mechanics the observer can consider any totality other than one including that observer in the act of performing the experiment; but the observer must always remain outside that system. Similarly, however great the totality of languages over which one generalizes, the language in which one does one's own generalizing must always lie outside the totality over which one generalizes. Thus, human subjectivity can never be thoroughly objectified within a complete, closed system; and this suggests that *there is no view from nowhere.*[19]

In setting forth his view of "pragmatic realism," Putnam steers a middle course between metaphysical realism and various interpretations of antirealism. Putnam rejects metaphysical realism, which he defines as the view that (1) the world consists of mind-independent objects; (2) there is exactly one true and complete description of the way the world is; and (3) truth involves some sort of correspondence between an independently existent world and a description of it.[20] The belief that science will one day describe the "way the world is" independent of any theory requires a leap of faith, however, for the whole history of science shows that it has always devised different mappings of the world. To believe otherwise is to rest on faith in a *future* science that will do something remarkably different from the science of the past and present.

Putnam also differs from many antirealists, for example, Bas van Fraassen, who make a strict demarcation between the *theoretical* entities of scientific theory, which are not observed (such as quarks, electromagnetic fields, and the charge of an electron) and observational entities, which are observed. Even observation—including technologically enhanced observation—is no absolute arbiter of objective reality, for it too is theory laden and subject to error. Rather than there being an absolute difference between theorizing and observing, we are faced with a smooth spectrum from a relatively theory-free perception, for example of the color red, to a relatively nonexperential conception, for example, superstring theory. Given such a spectrum, it is important to use a plurality of methods and theories to evoke multiple worlds for different purposes. All perceptual and conceptual knowledge gleaned in this way is always provisional and contextual. This fact does not imply that nothing exists prior to, and *in that sense* independently of, human experience, but does imply that the universe as we experience it does not exist independently of our perceptual and conceptual faculties, the operations of which are normally fused to varying extents.

According to James's radical empiricism, the whole range of perceived objects—from the macro-objects of commonsense experience to the minute objects perceived with scientific instruments—are accepted at face value, without attributing to any of them the property of absolute, or intrinsic, existence. Thus, our image of the world cannot be "justified" by anything but its success as judged by the interests and values that evolve and get modified at the same time and in interaction with our evolving image of the world. In James's words, *"True ideas are those that we can assimilate, validate, corroborate, and verify. False ideas are those that we cannot."*[21] On this theme he continues as follows.

> Truth lives for the most part on a credit system. Our thoughts and beliefs "pass," so long as nothing challenges them, just as bank-notes pass so long as nobody refuses them. But this all points to direct face-to-face verifications somewhere, without which the fabric of truth collapses like a financial system with no cash basis whatever.[22]

In establishing this criterion for validating hypotheses, James was actually corroborating the criterion already used by the empirical sciences. Scientific theories have never been validated by their correspondence to some independent, objective reality. Rather, science has always "bootstrapped" its way to more and more valid (or less and less flawed) theories. As science progresses, theories that previously seemed valid are discovered to be flawed, or are recognized as limiting cases of a more comprehensive theory, and are replaced by new conclusions or working hypotheses. In addition to the purely epistemic criteria of valid theories, there have always been pragmatic considerations; for scientists have long been concerned with the *usefulness* of any given theory.

Is the question of the validity of our perceptions simply a matter of our local epistemology and the standards of the time? Not if we accept a hypothesis that there is a level of human perception and rationality that is primary, in the sense that humans throughout history, in diverse cultures, experience and understand the world in common.[23] Thus, the meaning and reference of reasonableness and justification may be at least partially equated across changes in our epistemological paradigms from one culture and era to another.

If all true statements fall within a conventional/factual spectrum, as Putnam proposes, we are then presented with the challenge of placing specific assertions within a corresponding spectrum of subjectivity and objectivity. Statements such as "My own musical compositions are the most beautiful I have ever heard," "David is my best friend," and "My way of preparing escargot makes for an exceptionally delicious appetizer" may all be true for a specific individual at a specific time in his or her life, but their truthfulness may stop there. Other statements concerning laws, moral codes, music, art, and language usage may generally be true only for a specific human society and not for others. Further along the spectrum toward objectivity, statements concerning immediate perceptual experience, for example, "unrefined wool cloth has a coarse texture" and "ginger has a pungent fragrance," may be true for some species but not for others. Finally, some assertions may be true for all conscious beings, without reference to their kinds of sensory faculties, modes of cognition, or their locations in space or time. The laws of physics and mathematics are believed by some to fall into this category, while certain religious statements are believed by others to be true for everyone for all time.

Religious believers may assert that certain statements are true from God's own perspective, while advocates of scientific materialism may assert that scientific truths are valid without reference to any perspective whatsoever. Once again we are confronted with the transition from a God's-eye view to a view from nowhere. For all statements, however, ranging from the most personal and subjective to the most impersonal and objective, there is a deeply ingrained tendency to reify their truths. That is, while all our statements are contingent upon our cognitive faculties, these subjective elements are easily forgotten or overlooked; and we easily come to view our assertions concerning the objects of our knowledge and beliefs as being objectively true in an absolute sense.

Returning the Mind to Nature

A major incentive for James's formulation of radical empiricism was to reintroduce the mind, including sensory and mental phenomena, into nature, from which it had been divorced by the theologically motivated, mechanistic philosophy of Descartes. Modern science has given us every reason to conclude that sensory phenomena, such as colors, sounds, smells, and so on, do not exist independently in the objective, physical world. Rather, these are events that arise in dependence on outer phenomena such as electromagnetic radiation and on the inner workings of the brain. In the absence of a brain, as far as we know, there are no phenomena of visual forms, sound, smell, taste, or tactile sensations. Nevertheless, we experience the world around us as if it consists of these phenomena independently of our perceptions of them. While the world of our sensory experience appears to have an objective reality, in fact it is more like a dreamscape or a rainbow: it can be perceived, but the objects as they are perceived have no independent, objective existence. Likewise, the subjective mind itself may also be nothing more than a matrix of events arising in dependence on other events, much as a rainbow occurs as a result of the interplay of light and raindrops.

These hypotheses concerning the nature of the world of human experience and consciousness have been suggested by many cognitive scientists, but they generally adopt the view of scientific materialism of contrasting the unreal world of the mind with the real, objective world of physical science. The underlying metaphysical supposition here is that the real world conceived by science, beyond the veil of subjective appearances, exists independently of humans percepts and concepts. The problem with this belief is that the world of science is described in human languages, using terms drawn from our human senses. Indeed, the very concept of the "real external world" of everyday thinking rests exclusively on sense impressions. Even such abstract notions as electromagnetic fields, superstrings, gravity waves, and black holes are linked to our sensory experiences of fields, strings, waves, and holes. There is a fundamental difficulty here in trying to describe something that purportedly exists independently of our senses with terms based in our sensory experience; and there is a corresponding difficulty in trying to describe something that purportedly exists independently of our concepts solely in terms of human concepts. The latter difficulty is compounded by the fact that a diversity of mutually incompatible theories can often be formulated that equally account for the same body of experimental evidence and that yield identical predictions. This *problem of underdetermination* is common throughout physics. In such cases, empirical evidence alone simply cannot decide which competing theory is supposed to represent the objective, physical reality. Thus, choices are made on the basis of the scientist's subjective pragmatic, aesthetic, and metaphysical predilections.[24]

A recognition of this problem may have been the basis of the assertion by seventeenth-century advocates of organic philosophies that it is impossible to discover the intrinsic, purely objective mechanisms of experienced phenomena. Their goal was to identify and use the properties of observed phenomena through *firsthand experience*, while the aim of the proponents of mechanical philosophies was to identify their purely objective properties through *reason*. The problem of underdetermination suggests that the viewpoint of the organic philosophies may have been the more defensible one, for it, unlike the mechanical philosophies, did not ignore the role of subjectivity.

Despite the problem of underdetermination, there is no doubt that some physical theories are more successful than others, and none has been more successful than quantum theory. Moreover, among the entire range of scientific theories, none has more severely challenged the metaphysical assumptions of scientific materialism. Physicist John Bell has shown that quantum mechanics is incompatible with the very existence of an underlying reality resembling the observed world at the macroscopic level, with its separate physical parts linked only by causal dynamical relationships. According to the prevailing interpretation of quantum mechanics, matter is no longer viewed as the primary constituent of reality but is viewed rather as an "objective tendency" or "potentiality" within the quantum domain. A dominant theme of this view is the profound unity in nature underlying all that appears to be separate.[25]

According to most interpreters of quantum theory, reductionistic determinism is no longer viable, and some distinguished physicists think there are good reasons for believing that the implications of quantum theory cannot be understood without understanding consciousness.[26] Specifically, the act of measurement—by which quantum potentialities transform into physical reality—remains an unresolved problem in this field. Physicists do not know precisely what it is about measurement that allows it to take such a crucial role in the quantum world, but the role of conceptual designation may be central to the transition from potentiality to reality. Although there is no consensus on this issue, it is clear that particles can form nonseparable, entangled combinations, and for such states the traditional designation of individual particles by attributes fails; for only jointly conceived attributes can be said to exist.

To examine the problem of measurement in relation to consciousness, we may start with the simple question. Does a yardstick falling in the forest with no one present measure anything? Consider the hypothesis that it does if, and only if, someone has created and used that yardstick with the intention that it should measure something. If so, the measurement of a stretch of ground may take place as soon as the yardstick falls over, *regardless of whether anyone is there to actually perceive it at that time*. The conceptual designation of "yardstick" and "measurement" is not isolated to a specific place or time, so immediate, on-site perception of the yardstick falling to the ground is irrelevant. However, if a notched piece of wood falls to the ground without any such conceptual designations of "yardstick" and "measurement," then it cannot be said that a measurement takes place at any time. For yardsticks and measurements do not exist independently of their conceptual designations. And conceptual designations do not exist independently of consciousness.

The same is true of quantum measurements. In this context, the experimenter's determination of the moment of measurement is not arbitrary but is a conscious act; and without that conscious act, the moment of measurement never takes place, and the so-called probability wave does not collapse. Moreover, the conscious act of determining the time and nature of a measurement is "omni-temporal," in the sense that regardless of when that conscious act takes place, the measurement that is designated may occur in the past, present, or future. Analogously, physicist John Wheeler suggests that due to "acts of observer-participancy," physical reality may manifest not only now but back to the beginning of the universe. For example, as we conceive of and measure cosmic background radiation, we thereby "create" the Big Bang and the evolution of the universe as we presently understand it.[27] In this way we create the reality of human experience with the questions we ask. Wheeler concludes, "I do take 100 percent seriously the idea that the world is a figment of the imagination."[28]

In the context of the mind/body problem, the theory that immediately before a decision or intentional action, the physical state of the brain may determine only alternatives or potentialities is consistent with physical causation as it is understood in quantum theory and chaos theory. No physical law of nature would be violated if a nonphysical mind were to select one of these alternatives.[29] In any case, as Hilary Putnam argues, on purely pragmatic grounds more understanding is gleaned by taking into account mental causation than by dogmatically attributing all events solely to physical causation. That is, we limit our knowledge by ignoring such subjective causal factors as human desires and beliefs and confining ourselves solely to the objective, unconscious workings of the brain.[30]

The problem of causality within quantum mechanics may be deeply connected to the principle of symmetry as it pertains to the question of the relation between mind and matter. Most quantum theorists maintain that all the sufficient causes for a quantum event cannot, even in principle, be known at any given time. Generally speaking, if a hypothetical entity cannot be known *even in principle*, there are no grounds for positing its existence; for existence cannot be posited independently of the possibility of verification. Likewise, the existence of a verifying cognition cannot be established independently of an entity whose existence is verified. Thus, we are confronted with a symmetry between existence and cognitive verification.

Consider another kind of symmetry that may exist between causes and their effects. The complete set of sufficient causes for any phenomenon B is not present until the moment immediately preceding the occurrence of B. The notion of a set of sufficient causes A "immediately preceding" an effect B indicates that the lapse of time between A and B—regardless of its actual duration—is so brief that no other influences could intervene to prevent the occurrence of B. Thus, only in that moment immediately preceding B is there total certainty as to B's occurrence. Until then, there is a variable probability function only. If it turns out that B does not occur, no causes of B exist at any time; for no causes of B can be posited if B does not exist. Thus, the existence of the causes of B can be determined with certainty only upon the occurrence of B, that is, retrospectively. Thus, the set of causes A cannot be said to exist except as a probability until B occurs. Likewise the event B cannot be said to be a result of A until that complete set of causes occurs. Thus, we encounter another symmetry between the determination of the existence of A and B. In this way, a relationship of mutual contingency exists between causes and their effects. Other examples of such mutual contingency are found in the spatial relation between left and right, the temporal relation between before and after, and the cognitive relation between subject and object. Following this line of reasoning, it appears that not only are effects functions of their causes but causes are functions of their effects.

In quantum mechanics, before a quantum event actually takes place, one can speak of only potentialities or probabilities. But the very existence of a probability can be posited only within the conceptual framework of (1) anticipating an effect that does not yet (and may never) exist and (2) identifying causes of that effect that may not exist. The strong tendency to reify the existence of elementary particles also extends to probabilities. But in this regard Werner Heisenberg cautions that "[i]f one wants to give an accurate description of the elementary particle . . . the only thing which can be written down as description is a probability function. But then one sees that not even the quality of being . . . belongs to what is described."[31]

For this reason Niels Bohr, another of the principal architects of quantum mechanics, declared that "[a]n independent reality, in the ordinary physical sense, can neither be ascribed to the phenomena nor to the agencies of observation."[32]

The existence of quantum entities such as photons cannot be posited independently of the agencies of observation; nor are they merely artifacts of the system of measurement or arbitrary figments of the imagination of the observer. In this regard, the subjective agency of observation and the observed object are entwined in two ways: (1) the observed object is invariably disturbed by the observation of it, and (2) the observer brings specific questions and a conceptual framework to the observation. Once a measurement of a quantum event is made, it determines both the past and future attributes of the event, though it doesn't causally affect the past. Large-scale aggregations of quantum entities, such as spatially dimensionless particles, consist entirely of interactions, or relationships. Thus, the physical universe actually consists of nothing more than dependently related events, as opposed to real, independent, objective entities that may or may not enter into relationships. Independent particles of matter can no longer be posited as the fundamental building blocks of the physical world. Moreover, even if such independent particles were to be out there in the objective universe, they could never be detected and could therefore not be posited to exist. For this reason, it makes no sense to speak of anything existing independently of everything else.

Over the past four hundred years, physics has repeatedly progressed from assumptions of asymmetry to principles of symmetry and from assertions of absolute entities to relative events. No longer are space and time viewed as independent, absolute realities; rather space and time are now seen as properties of space/time; mass and energy are both seen as interchangeable properties of matter (which itself is now regarded as consisting of principles of symmetry rather than little bits of stuff); and quantum relativity proposes that there is a comparable relation between potentiality and actuality. Historically speaking, the assertion of something that *is acted upon* but does not *act* is a sign of a degenerate theory. But just such a theory is still maintained by many scientific materialists concerning the nature of subjective mental events: they are acted upon by the body, but they exert no influences on the body. If one surveys the references to mind/body interactions in philosophical, scientific, and medical literature over the past century, the overwhelming emphasis is on the body's influence on the mind, with relatively very little attention paid to the mind's influence on the body. In light of the history of physics, this is a clear indication of a degenerate theory.

The broader context of this asymmetry includes the premises that (1) objective and subjective, (2) physical and mental, and (3) outer and inner phenomena are absolutely different, with the causal influences going from the former to the latter but not vice versa. In short, these lingering absolute dualities, which are our legacy from Descartes, have long outworn their credibility in light of modern physics. The time is surely ripe to explore even more radical versions of relativity and symmetry, in which one may speak of subjects and objects as being mutually entwined properties of a subject/object field. Thus, the absolute distinctions between mental and physical, conceiver and conceived object, and outer and inner phenomena may be absorbed into more comprehensive theories that include each of these dualistic aspects as parts of a greater whole. As mentioned earlier, consciousness stands alone today as a reality that has resisted all satisfactory explanation within the parameters of the principles of scientific materialism; and the assumed asymmetry between consciousness and the objects of consciousness is perhaps the single most flagrant, lingering, degenerate theory in modern science. We now seem to be faced with two options: discard consciousness or discard scientific materialism.

The various branches of modern science have developed on the model of physics, and scientific materialists have traditionally looked to physics for validation of their beliefs. But now quantum theory, the most successful of all physical theories, appears to be the single greatest threat to the credibility of the metaphysical assumptions of scientific materialism. Specifically, according to quantum theory, the brain as a real, purely objective composite of particles of matter can no longer be deemed to exist; so the reductionist attempt to reduce all mental phenomena to this classical conception of matter is radically undermined.

To discuss the mind/brain problem today without taking into account the implications of quantum theory is like discussing the movements of the planets without taking into account the Copernican Revolution. It is reported that some of Galileo's clerical opponents were loathe to gaze through his telescope to take a closer look at the planets, sun, and moon for fear that what they saw would violate their beliefs. In a similar fashion, many cognitive scientists are loathe to observe their own minds, for the principles of scientific materialism deny that such observation is possible; or even if it is, the phenomena observed introspectively must be misleading or nonexistent. What new avenues of scientific inquiry might open up if we were to challenge this dogmatic injunction against the firsthand, empirical investigation of mental phenomena?

$$\circ \quad \circ \quad \circ \quad 4$$

OBSERVING THE MIND

Like the Copernican shift from a geocentric to a heliocentric view of the solar system, the shift from scientific materialism to radical empiricism entails a shift from a matter-centered concept of reality to a holistic view of mental and physical phenomena as dependently related events. In terms of scientific materialism, there is one taboo against scientific inquiry into subjective mental phenomena; and there is another taboo against allowing one's own subjective perspective to taint any scientific research. Thus, first-person, introspective inquiry into the mind is doubly taboo. This point is illustrated by events in a 1994 conference sponsored by the Royal Society in London entitled "Consciousness—Its Place in Contemporary Science." This meeting revealed a remarkable consensus among the speakers that science understands none of the central aspects of consciousness—what it is, how it evolved, how it is generated by the brain, or even what it is for. The paradox confronting the participants was that from the first-person perspective, consciousness is a prime irreducible datum, but from the third-person scientific perspective there is no way of investigating it directly. That is, brain research tells us nothing about why neural processes should give rise to mental experiences of any kind. However, when one participant suggested that research into consciousness must include the first-person perspective, a number of his colleagues expressed consternation. In their eyes avoiding the taboo of subjectivity and remaining ignorant of consciousness was apparently preferable to breaking that taboo and opening the possibility of fresh avenues of understanding.[1]

Introspection is given only marginal treatment in modern psychology textbooks, and in both psychology and the brain sciences, theorizing about the nature of introspection is at a rudimentary stage in comparison with other types of cognition. A central reason for this may be, as philosopher Daniel Dennett points out, that introspection, together with consciousness itself, are features of the mind that are most resistant to absorption into the mechanistic picture of science.[2]

A Historical Sketch of Introspection

If mental states exist solely as first-person, subjective phenomena, as suggested by everyday experience, the first-person point of view should certainly be primary; and we should let this subject matter dictate our research methods, rather than the converse. This implies the use of introspection as a primary method of cognitive science, but scientific resistance to this proposal is strong. One legitimate reason for this aversion is that introspection has already been tried out by philosophers and psychologists, and its failure to produce reliable scientific knowledge is a historical fact. With this objection in mind, here is a brief review of the history of introspection in the West.

Augustine was among the first Western thinkers to write on the topic of the firsthand observation of mental phenomena. In his treatise *The Free Choice of the Will*, he discusses the existence of an "inner sense," or mental perception, that functions distinctly from the five physical senses. While the eyes perceive colors but not the phenomenon of seeing and the ears hear sounds but not the phenomenon of hearing, this inner sense perceives both the objects of the five outer senses and those sense operations themselves. Such mental perception, he asserts, is a kind of arbitrator, or judge, of the external senses, for it decides what is and is not sufficient for the various outer senses. For example, it observes whether or not one has seen enough of an object, for it is aware of pleasure and pain. He admits to uncertainty as to whether the inner sense also perceives itself. Augustine distinguishes between this inner sense and reason. The outer and inner senses *perceive* their respective objects and "report" them to reason; but reason alone has the capacity of *knowing*. The inner sense does not truly understand, for it lacks intellect; while reason is more powerful than any of the other senses and comprehends itself as reason.[3]

At the end of the medieval era, Descartes, discussed introspection, as did Augustine, as an element of his proof of the existence of God. In his *Meditations*, he claims that with the "natural light of the mind" whatever is perceived distinctly and clearly is necessarily true. Errors in introspection arise only when one judges that the ideas inside one's mind resemble things outside the mind or are modeled on them. The more precisely one examines the contents of the mind, without referring them to anything else, the less is there any room for error.[4]

The word "introspection" first appeared in the second half of the seventeenth century, and the golden age of this mode of inquiry lasted from then until the first decade of the twentieth century. Throughout most of that period, introspection thrived in a secular, philosophical context; but by the closing decades of the nineteenth century, science finally turned its attention to the empirical study of the mind, and introspection was chosen as an important means of accumulating scientific data concerning mental phenomena.

Perhaps no one played a more influential role in the initial development of this "introspectionist school" than the German physiologist and psychologist Wilhelm Wundt. The challenge he faced was to present a model of the introspective observation of subjective, mental phenomena so that it appeared akin to the well-established, scientific modes of extraspective observation of objective physical phenomena. His response was to try to order and control the *external* conditions of introspection by having subjects sit still and confront simple perceptual stimuli, such as a green triangle, and to report according to well-defined rules. Such visual stimuli were presented for very brief and accurately timed periods with a "tachistoscope," and the reaction times between the presentation of the stimulus and the introspective report on the ensuing sensation were recorded with a metronome or chronograph. Wundt believed that scientific data could be obtained only from subjects who had been put through this routine at least ten thousand times.

Thus, the practice of introspection was distanced from the philosophical introspection of John Locke and even further removed from the contemplative introspection of Augustine and transformed into a repetitious, robotlike performance that seemed to Wundt to fulfill the criteria for scientific observation. The contemplative, philosophical, and simple everyday practices of introspection were deemed hopelessly unscientific; introspection was thought to provide reliable, scientific data only through such external restraints.

A similar rationale determined that subjects would introspectively focus on simple perceptual stimuli, for Wundt believed that more complex mental phenomena, such as thoughts, volitions, and feelings, were not sufficiently amenable to experimental control to be objects of scientific inner perception. Such inner observation was so contrived and hedged in with rules and regulations that to the uninitiated layperson it looked like an esoteric rite, far removed from anyone's commonsense experience of introspection.

Already in the 1880s the introspection-centered approach to psychology began to decline. American students trained in Germany returned to establish psychology laboratories in major American universities; and graduate schools soon sprang up that were exclusively modeled after the German doctoral system, in which the professional boundaries between psychology, philosophy, medicine, and other disciplines were strictly demarcated. In this academic atmosphere, psychology became determined at all costs to associate itself with the physical sciences. Thus, methodology took precedence over subject matter, and the ideal of generating objective, scientific data displaced the significance of the individual.[5]

In his illuminating essay "The History of Introspection Reconsidered," focusing on academic psychology during the period 1880–1914, Kurt Danziger concludes that the total rejection in principle of introspection was not a rational conclusion in the light of the problems that the method encountered. Rather, it was due to a *shift of interests* among psychologists, especially in America. "Such interests," he points out, "redefine the goals of psychological research and hence produce a re-selection of the methods needed to achieve these goals. Introspection was less a victim of its intrinsic problems than a casualty of historical forces far bigger than itself."[6] Once again the truth of James's attentional reality principle is illustrated: the knowledge that could be provided by means of introspection no longer held the *interest* of modern psychologists; as a result they no longer *attended* to it; and thus introspection lost its place in the psychological understanding of the mind.

The external reason for the failure of introspectionism was the rising influence of positivism in all sectors of science, as well as the humanities. The chief internal reason for its collapse was the fact that the word of the subject was the final authority with regard to mental data; and when different subjects' reports turned out to be mutually incompatible, the introspectionist movement found itself in a theoretical quandary from which it was never able to extricate itself.

For all that movement's efforts to conform to the scientific tradition by reducing introspection to a mechanistic mode of detecting primitive sorts of mental phenomena, those methods proved incapable of producing reliable psychological data; and the school of introspection was soon superseded by behaviorism. But behaviorism never accomplished its goal of translating Cartesian mentalist accounts into behavioral ones, nor did it ever cope successfully with the "problem of privacy" in general or the nature of introspection in particular.

In the latter half of the twentieth century, behaviorism was supplanted by neuroscientific methods of investigating the mind. Functionalist accounts have been very prevalent in these recent brain-centered theories of the mind, but it is not clear what, if any, information they provide as to the real nature of what humans do when we introspect. Indeed, after behaviorism, mainstream theoretical psychology and philosophy have had little to say about the nature of introspection. While theories have appeared that depict introspection as a literal reporting on discrete brain states or processes, there is little or no scientific basis for such views. The mental terminology normally used when describing such introspective reporting bears only a very indirect relationship to actual brain processes. Moreover, if introspection in this sense were to provide us with immediate access to and knowledge of the brain, it would yield knowledge about neuronal firings, the state of the neuron-protecting glial cells, and the intricacies of cerebral processes and states; but this has not proven to be the case.

In both psychology and the brain sciences, theorizing about the nature of introspection remains at a primitive level in comparison with theorizing about other cognitive processes such as perception and memory. And while introspection continues to be dismissed in psychology as a means of studying mental phenomena, it is still marginally retained as a crude, unscientific appendage to serious scientific research.

Arguments Against Introspection

Ideological Objections to Introspectionism

The introspectionist school met its demise as a result of both ideological and pragmatic, scientific problems. One ideological objection was that the principle of objectivism demands of scientific observation a kind of independence of subject and object that is impossible in introspection. Wundt acknowledged that subjective events can be internally observed, but he argued that this does not imply that such events are observable in any *scientific* sense. To get around this problem, he advocated a form of "internal perception" (*innere Warnnehmung*), the conditions of which were manipulated so that they approximated the conditions of external perception. Subjects trained in such "internal" perception made observations and "judgment-free" reports on their perceptions; while most mental phenomena, including thoughts and complex feelings, were excluded from introspective study.

The kind of independence of subject and object that Wundt and many of his contemporaries idealized can be traced back to the Scholastic era. As William James pointed out,

> in scholastic theism we find truth already instituted and established without our help, complete apart from our knowing; and the most we can do is to acknowledge it passively and adhere to it, although such adhesion as ours can make no jot of difference to what is adhered to.[7]

The French philosopher Émile Boutroux, following James, argued that from the philosophical standpoint no absolute divisions between the subjective and objective are given of the sort that science imagines for its convenience. "Continuity," he declared, "is the irreducible law of Nature."[8] Boutroux, however, took this a step further:

> Here we encounter the real problem which is at the heart of this discussion: is there no other experience than that which the duality of a subject and an object implies? May not this experience, belonging to distinct consciousness and to science, be derivative and artificial, in comparison with that primary and genuine experience which is truly one with life and reality?[9]

If this is the case, the absence of any absolute demarcation between subject and object in the process of introspection might allow for "a more primary and genuine experience" than is possible by means of extraspective, scientific observation. In other words, introspection may provide the sole access to nondual knowledge that is truly *natural*, while dualistic scientific knowledge of the world may be fundamentally *unnatural*. The oddity of this hypothesis stems from our long habituation with Scholastic realism—a metaphysical position that modern science has largely assumed through the twentieth century.

A great irony regarding the violation of objectivism is that the academic psychologists who rejected introspectionism in favor of behaviorism were of the same generation as the pioneers of quantum mechanics. In this revolutionary branch of modern physics it is common knowledge that extremely minute physical events cannot be studied independently of the mode of observation; and scientists do not know whether quantum entities even exist independently of their measurements. This theme is specifically addressed in the well-known Heisenberg Uncertainty Principle.[10] The participatory nature of scientific observation in quantum mechanics has given rise to a great deal of fascinating debate among physicists and philosophers; while the participatory nature of introspective observation in psychology has been taken as grounds for rejecting the very possibility of such scientific observation.

A second metaphysical objection to introspection is based on the premise that mental events in general, and all causally efficacious mental processes in particular, are unconscious and therefore inaccessible in principle to introspective observation. This theory can be traced back to Leibniz and Kant. While Leibniz denied the equation of the mind and consciousness, implying that the nature and constitution of the mind may not be accessible to consciousness, Kant went further in declaring that introspection is limited to the world of psychological appearances, which has little relevance to the real constitution of the human mind. Thus, the true basis of mental phenomena—namely, the subject that knows, wills, and judges—is inaccessible to inner experience. For this reason, the description of the subjective mind must remain on a purely anecdotal level and cannot achieve the status of a science.[11]

A modern, materialistic reinterpretation of this view asserts that mental events in general, and all causally efficacious mental processes in particular, are unconscious, for they are actually brain states that can be studied solely by objective, scientific means. Some cognitive scientists fly in the face of experience by arguing that *no* activity of the mind is ever conscious![12] Thus, the real constitution of the mind, which Kant assumed to reside in the inaccessible realm of noumena, is now thought to lie hidden in the brain. And all first-person accounts of mental phenomena are condemned to the status of folk psychology, with no possibility of their rising to the standards of empirical science.

If one were to apply the same reasoning to the extraspective observation of the physical world, one would be led to the conclusion that visual observation—either with or without the aid of such tools as telescopes and microscopes—is limited to the world of physical appearances, which are profoundly misrepresentative of the real constitution of the physical universe. But while this dualistic construct of phenomena versus noumena has done little to hamper the physical sciences, it has contributed to the stifling of introspection as a means to exploring the mind.

The current belief that all mental processes are unconscious is so obviously contrary to experience that it can be regarded simply as a symptom of the metaphysical miasma induced by overexposure to scientific materialism. On the other hand, it is not so easy to dismiss the hypothesis that mental phenomena, or at least those that are accessible to introspection, are devoid of causal efficacy. This expression of the closure principle presumably stems from the assumption that causality necessarily entails a mediating *mechanism*. Thus, advocates of the notion that mental phenomena exert causal influences among themselves and upon the body are challenged to produce a mechanism by which such influence might be exerted. This metaphysical burden has been carried by mind/matter dualists at least since Descartes; and their failure to produce a mechanism by which the mind influences matter has been central to the decline of Cartesian mind/body dualism. However, modern science, adhering to the principles of scientific materialism, has *equally failed to explain what it is about the brain that allows it to create and influence conscious mental events*. In the meantime, the demand for a mechanistic explanation of causality has been long rejected in various fields of physics, including electromagnetism and quantum mechanics.

In the history of physics, the phenomenological study of dynamics preceded the theoretical formulation of mechanics. For example, the empirical observations of the movements of planets and terrestrial objects performed by Tycho Brahe, Kepler, and Galileo were necessary before Newton could stand on the shoulders of these giants and formulate his mechanical laws of nature. But when it comes to consciousness, scientists are taking the opposite approach. They are trying instead to formulate mechanical theories of consciousness without *ever* relying upon precise, firsthand observations of states of consciousness themselves. This approach is far more analogous to that of medieval astronomers than that of the founders of the Scientific Revolution.

An alternative, phenomenological interpretation of causality that is most appropriately applied to mental causation asserts simply: if a set of one or more events A precedes an event B, and B does not occur without the prior occurrence of A, then A is said to cause B. This concept of causation can be put to the test in individual cases only retrospectively; but this is the way we normally conclude that one event caused another. With this less metaphysically burdened concept of causality, it becomes perfectly obvious that mental phenomena do act as causes of subsequent mental and physical events. It is equally obvious that physical phenomena act as causes of subsequent mental and physical events. These facts must be acknowledged *regardless of whether one has found a mechanism by which such causality is made possible*.

Cartesian dualists reify both objective physical processes and subjective mental processes—taking both types of phenomena as inherently existing, independent substances—and they have never provided a satisfactory explanation for how these two different types of substances can interact. Philosophical idealists who reify the mind by asserting it as an inherently existing, independent entity—while maintaining that objective phenomena are mere epiphenomena of the mind—have never provided a satisfactory explanation for how physical epiphenomena can influence the mind. Philosophical materialists who reify matter by asserting it to be an inherently existing, independent substance—while maintaining that subjective mental processes are mere epiphenomena of matter—have never provided a satisfactory explanation of how mental epiphenomena can influence the body. The common error in all three of these philosophical positions is reification, which makes it impossible to construct compelling theories accounting for interrelationships among reified entities of any kind.

The causal interaction between mind and matter becomes an insoluble problem as soon as one defines either as an independent, inherently existent substance. This is equally true of interactions among any disparate phenomena, such as particles and fields, within the material world. Moreover, if one were to postulate the existence of elementary particles as independently existent bits of matter, not only would physicists have no way of knowing of their existence, such particles would have no way of interacting among themselves! As soon as one begins to understand subjective and objective, mental and physical phenomena as *relational* instead of *substantive*, the causal interactions between mind and matter become no more problematic than such interactions among mental phenomena and among physical phenomena. But the notion of a reified causal *mechanism* may no longer be useful in any of these domains.

Since the mind alone perceives both mental and physical events, as well as the relations between them, introspection should naturally play a vital role in determining such causal interactions. To be sure, the hypothesis that introspection may provide knowledge of causal mental phenomena does not necessarily imply that all mental influences are immediately accessible to inward observation. However, some individuals who are particularly adept at introspection, such as William James, may observe an exceptionally broad range of causal mental influences. If so, it seems plausible that repressed, unconscious, and preconscious mental processes that simply happen to be unconscious may, with training, be brought into the light of introspective awareness. Such training might also alleviate the problem of introspection actually destroying, or, at best, grossly distorting its object. Although the participatory nature of introspective observation may well be inescapable, it may be possible to refine this process so that distortion is minimized. This hypothesis can be tested only through experience.

Conversely, if one is indoctrinated into a belief system that denies the possibility or value of introspection, this may cause one's introspective abilities to atrophy. If so, scientific materialism's assessment of introspection may well have been conceived by individuals with impaired introspective abilities; and its adoption by others may result in the atrophying of their ability to observe their own mental processes. Thus, while children normally develop the ability of introspection by the age of eight, their later indoctrination into the principles of scientific materialism may actually cause them to revert to a preadolescent state of psychological immaturity. The denial of first-person awareness of conscious states thereby becomes a self-fulfilling prophecy.

Scientific Criticisms of Introspectionism

Critics of the scientific use of introspection have raised the legitimate problem that when the introspecting subject is compelled to reply to the questions of the experimenter, this not only biases the observations and responses but also carries the implicit message to the subject that all the questions are answerable. To place this problem in a broader context, consider the fact that all observations—scientific and otherwise—are theory laden; that is, they are all colored by the types of questions we pose in our acquisition of empirical data. Nowhere is this more evident in the physical sciences than in the field of quantum mechanics. Werner Heisenberg, for example, comments that "[w]hat we observe is not nature in itself but nature exposed to our method of questioning."[13] Einstein comments in a similar vein that "on principle, it is quite wrong to try founding a theory on observable magnitudes alone. In reality the very opposite happens. It is the theory which decides what we can observe."[14]

Since neither behavioral science nor brain science has direct access to any mental phenomena, their accounts of the nature and constitution of such phenomena may be at least as biased by culturally conditioned conceptual frameworks as are the accounts drawn from introspection. Furthermore, in the course of seeking knowledge of mental states by means of introspection, it is well to bear in mind that various types of experience may be impossible to communicate to those who have not themselves experienced them.

Another closely related scientific objection is that when the words in which the experimental subject describes his or her experiences do not induce in the experimenter corresponding experiences of his or her own, a specific interpretation, hence a scientific evaluation, of such introspective reports is impossible.[15] Like the previous objections, this problem is not confined to introspection. For example, I know the difference in taste between wine and wine vinegar; but I would have difficulty in adequately describing the difference between these tastes to a person who has tasted one and not the other. I would have all the more difficulty in describing it to a person who has tasted neither. On the other hand, wine connoisseurs are able to describe among themselves subtle differences among vintages. I cannot detect those differences, nor do I really understand what they are talking about, even though they are speaking in clear English. Similarly, masters at introspection may be able to discuss certain experiences among themselves, while others listening in could literally not make sense of their conversation. Such communication may not be different in principle from other instances of privileged conversation that commonly occur among highly trained mathematicians, scientists, musicians, and so on.

A solution to this problem of communication is that the experimenter be at least *as* experienced, if not *more* experienced, in introspection than the subject. The mediation of such an individual who has greater experience has been called a second-person perspective. This process appears in the natural sciences when a researcher seeks mediation from a more experienced tutor in attempting to improve his or her skill as a scientist, but such second-person mediation generally disappears by the time the young scientist publishes an article in a scientific journal. For the study of the mind, there is no reason why the experimenter could not be his or her own subject, though there may be benefits to conducting such research under the guidance of an even more experienced practitioner. The latter option is more akin to the relation between contemplative mentor and disciple, which is a far cry from the orthodox paradigm of psychological research.

Another objection raised against the scientific use of introspection, particularly of the sort promoted by Wundt, is that such introspection is so artificial and contrived that it bears no relevance to everyday introspection. As noted earlier, when introspection was engineered so as to conform as closely as possible to extraspective, scientific observation, it could no longer be used to inquire into any but the most primitive of human cognitions, while the higher functions of thought and feeling were ignored. Such research failed to capture the interest of the public at large (for obvious reasons); and when the pioneers of behaviorism presented another mode of research that seemed to exclude subjectivity altogether, introspection was abandoned. Another response that might have been pursued if the pressure of scientific materialism had not been so dominant is to develop ways of refining everyday introspection to allow for observation of a broad range of mental states with increasing reliability and precision.

Another problem with the possibility of the scientific use of introspection is that science demands that when an experiment is run twice with the same initial conditions, you should get the same results. But such predictable, uniform results have not been yielded by introspective observation. One major reason for this is the immense complexity of the brain and mind. With such complexity, you can't expect to establish exactly the same initial conditions for two different individuals or even for one individual at two different times. But the fact that something is exceptionally complex simply means that the standards of scientific observation have to be adjusted to that reality; one should not exclude the possibility of one kind of scientific observation simply because it doesn't conform to other types of scientific observation.

Even when two subjects do report very similar experiences based on their introspective observations, how can we ever confirm that either one has actually had the reported experience? The privacy of introspection seems to make any objective confirmation impossible. This is just as problematic as trying to determine whether another person has actually understood a mathematical proof. Did the "proof" actually prove anything to that person? All we have is his or her subjective report. We might infer that he or she did understand the proof based on his or her own subsequent work in mathematics. But if there are no behavioral skills acquired by introspective reporting, there is little to go on except faith in the subject's introspective abilities and integrity.

This problem, too, can be extended to scientific research in general. When any scientific discovery is reported, how do we know that the findings are sound unless we have the expertise and opportunity to do the experiment or research ourselves? Even if we know the data are replicated in many other researchers' laboratories, how do we know they are not all wrong because of a common error, as in the case of the repeated confirmation of the existence of N-rays? Similarly, in the field of mathematics, how do we know another's mathematical analysis or proof is sound unless we have the ability and time to confirm it for ourselves? In all such cases, scientists, mathematicians, and the general public simply place their faith in the ability and integrity of the researchers; and we are all encouraged that our faith is well placed when we observe the pragmatic benefits of technology yielded by such research. Thus, if introspection is ever to have a place in scientific research, it is reasonable to demand that it produce not only firsthand accounts of its findings but observable, practical benefits as well. It remains to be seen just what those benefits might entail.

Modern Philosophical Refutations of Introspection

A modern philosophical analysis of introspection appears in William Lyons's work *The Disappearance of Introspection*, in which he argues that genuine introspection—in the sense of a metacognition, or internal monitoring, of such mental phenomena as perception, memory, imagination, thinking, and so on—never takes place at all. In his view we are never able to observe any of our own conscious mental processes, nor do we view mental copies of earlier experiences. Thus, all accounts of introspection as a form of monitoring, inspecting, scanning, or immediate retrieval of data with respect to cognitive processes, he says, constitute a "myth of our culture, an invention of our 'folk psychology.' The alleged introspection of perception is another sort of myth, by and large a concoction of psychologists and, especially, philosophers."[16]

In presenting his argument against the very existence of introspection, Lyons correctly asserts that we commonly have an unfounded certainty about the reliability of our internal observations. One reason for this is that too much occurs mentally for us to attend to all of it; and we may not focus on, or recognize, the most influential mental processes. Indeed, some of the mental processes that largely govern our behavior may not be immediately accessible to introspection. Our conscious experiential life, Lyons insists, is made up solely of the exercise of our senses and perceptual memory and imagination. All other mental processes play no part in our experiences, for they lie "hidden in the dark, silent labyrinth of the brain."[17] As a result, this domain of experience "will remain more or less unknown to us until neuroscientists gradually unfold its mysteries to us."[18]

This faith placed in future neuroscientific knowledge might well come as a surprise since Lyons acknowledges that even with immediate access to and knowledge of the brain, one may learn a great deal about cerebral processes and states but nothing about our mental life. Paradoxically, Lyons denies that introspection provides immediate access to any mental phenomena, concedes that the study of the brain fails to account for these subjective events, then places his faith in nonexistent, future neuroscience to come to the rescue!

In defense of his thesis that accounts of introspection are nothing more than inventions of folk psychology, Lyons cites empirical evidence indicating that the ability, and some recognition of the ability, to do what goes under the label of "introspection" is first acquired by children around the age of eight.[19] Children, he says, begin to develop their folk psychology by abstracting the concept of the mind from dreaming, imagination, and internally heard speech. This concept is then reified, and the mind comes to be viewed as the source of these activities; and it is from this conceptual framework that children (and adults) communicate with others about their inner cognitive life and come to understand the inner cognitive lives of others. Lyons quite plausibly views this evidence as indicating that by the age of eight, children in our society have been indoctrinated into the Western concepts of "introspection" and "folk psychological" accounts of the "mind" that can be inwardly observed. Directly as a result of this cultural conditioning, they pick up the misguided sense that they are actually able to observe at least some of their mental processes.

Consider, however, another way of viewing the same evidence. The tendency of the psychologically immature mind may be to attend solely to external, objective events; but by the age of eight, children's minds mature to the point at which they are able to focus inwardly on at least some of their own subjective processes. As noted previously, the ability to make scientific observations with a microscope is also an acquired skill, which depends on sophisticated theoretical training. Without such training, scientifically one is practically blind. One needs to learn what to look for, and one needs to learn how to recognize it when it is before one's eyes. This is simply a fact of modern scientific research, and there are few who dismiss this as simply a case of cultural indoctrination or folk psychology. If these two cases are comparable, the ability to introspect may be a sign of psychological maturation rather than cultural indoctrination. If so, indoctrination into scientific materialism may actually impede, or counteract, such maturation.

Lyons also rejects introspection as a form of monitoring conscious mental processes by citing scientific evidence that simultaneous performance of two attentive acts of cognition rarely, if ever, occurs.[20] This empirical conclusion, he asserts, casts doubt on the possibility of simultaneously attending to an object of consciousness and to the subjective consciousness of that object. Such research, however, tests the hypothesis of simultaneous attention to dissimilar objects in different sense fields, such as a visual stimulus and an auditory stimulus. While a visual object and the consciousness of that object are dissimilar to some extent, they are certainly more closely related than a visual and an auditory object; so it is not entirely clear that this research has a direct bearing on introspection. Moreover, the notion that introspection entails a split in the attention is not the only possible interpretation. William James, for example, views introspection as a retrospection of a mental event held in short-term memory. Thus, the mental events that are so observed are enveloped within, or are processed by, a conscious process but are not themselves conscious.

Another of Lyons's arguments against introspection is that while extraspective perception is based on the organs of sight and the other senses, and these can be checked easily to see that they are in working order, this is not the case with introspection. There is no agreement, he observes, as to the necessary organs of introspection, let alone agreement as to how they might be checked to see whether they are in working order. This point should come as no surprise, however, in light of the fact that most of the higher order mental functions are little understood by contemporary neurophysiology and there is no cogent neuroscientific understanding of the production of consciousness. Furthermore, as Lyons acknowledges, within the brain sciences there is presently a general consensus that, while there is evidence for some sort of localization, there are no precise and different locations or even different sorts of brain tissue corresponding to different thoughts, sensations, or even different types of thoughts or sensations. Thus, the fact that no organs of introspection have been discovered can hardly be counted as grounds for rejecting its existence.

In all these arguments against the scientific use of introspection, a double standard is used for mental and physical phenomena. Many of the objections to introspection apply equally to extraspective scientific observation, where they are widely known and accepted. In all the rest of the sciences there are no injunctions against careful, direct observation of the phenomena under investigation, whenever this is possible. And such phenomena are commonly to be examined as objectively—that is, with as much precision and freedom from bias—as possible. But when it comes to mental phenomena—running up against the taboo of subjectivity—such observation is rejected in principle. It seems that scientific materialists would rather ignore mental phenomena than look at them, and they would rather impair our natural introspective abilities than refine them so that they may rise to the standards to scientific inquiry. A striking exception to this rule is found in the writings of William James on the nature and value of introspection for the study of the mind.

William James's Introspective Strategy

Before the advent of modern psychology, the role of the mind in nature was already marginalized, a position that scientific materialism has turned into a dogmatic principle. After enormous, prolonged expenditure of human effort and material resources, and by developing increasingly sophisticated technology, science has provided us with cogent, empirically based, scientific conclusions concerning the origins, constitution, and causal influence of stars and other natural phenomena millions of light-years away. But in the meantime, questions concerning the origin, constitution, and causal efficacy of mental phenomena remain in the domain of philosophers, for science has failed to supply compelling answers. Indeed, William James suggests that a topic remains a problem of philosophy only until it has been understood by scientific means, at which point it is taken out of the hands of philosophers.[21] Judging by the plethora of contemporary philosophical works on the mind and consciousness, science has not yet grappled successfully with human subjectivity.

A distinguishing characteristic of science is that it has developed in close cooperation with the development of tools and methodologies for making ever more penetrating and reliable observations of physical phenomena. This is an important factor distinguishing science from philosophy. However, scientists have made no similar progress since the time of Aristotle in developing tools or methodologies for examining mental phenomena directly. In this regard, the present situation of scientific understanding of mental phenomena may be likened to the late medieval Scholastics' confrontation with the external world of nature. The chief obstacle that hindered their pursuit of understanding may have been not simply adherence to the mistaken theories of Aristotle, but the lack of alternative modes of empirical and theoretical inquiry. To put it more bluntly, the major problem may have been the active suppression of alternative modes of research by the dominant ideology of the time.

In the twentieth century, scientific materialism was the ideology that suppressed modes of inquiry into mental phenomena that do not conform to its principles. Modern science began, with the Copernican Revolution, by displacing humanity from the center of the natural world, but scientific materialism has gone to the extreme of denying human subjectivity any place at all in the natural world. This dogma would rather deny the existence of introspection, or at least marginalize its significance, than acknowledge that, four hundred years after the Scientific Revolution, we still have no scientific means of exploring consciousness directly. In this regard, we are right now in a dark age; but the extent of our ignorance of mental phenomena is obscured by the extraordinary progress that has been made in the physical sciences, including modern neuroscience.

More than a century ago, James set forth a research strategy for a science of the mind that is at all times person-centered, but his proposals have rarely been adopted because of their incompatibility with scientific materialism. What made him all the more threatening was that he was concerned not only with the details and methodology of the empirical study of the mind but with examining the philosophical assumptions of scientific materialism itself.[22] A central aim of James's view of psychology was to reestablish the presence of mental phenomena in the natural world. In making this point, he was simultaneously rejecting the Cartesian, theological notion of all activities of the human soul occurring outside of nature and the materialist premise that subjective states either do not exist or else must be equivalent to objective, physical processes. Pointing out that the psychology of his day was hardly more developed than physics before Galileo,[23] James envisioned the possibility of psychology discovering how individuals could control the conditions of their own mentation—an achievement that would, he thought, dwarf the discoveries of the other sciences.

To open the way to explore these issues scientifically, James presented psychology as the study of subjective mental phenomena and their relations to their objects, to the brain, and to the rest of the world; and he argued that introspective observation is always the first and foremost method by which to study these issues.[24] But introspective study, he argued, must be complemented with comparative research, such as studying the behavior of animals, and experimentation, such as experimental brain science. He said that while introspection is no sure guide to truths about our mental states, as he freely acknowledged, it may also not be as thoroughly misleading as it is commonly presented to be.

It is easy to respond to James's proposal by pointing out that introspection has already been tried by the introspectionist school of psychology and failed miserably. However, the type of tedious, automatonlike, internal observation that was used in the introspectionist school was so boring and unfruitful that even James dissociated himself from such experimental research. These early introspectionists, in their zeal to acquire objective, scientific knowledge of subjective mental states, treated their human subjects like primitive laboratory animals. Their objective solution to the fallibility of introspection was to apply external, artificial constraints on their introspecting subjects, thereby reducing the sophisticated, human ability of introspection to a primitive, robotlike process of internal monitoring. After academic psychology treated human subjects like laboratory animals and found that they did not live down to that standard, it should come as no surprise that it then shifted its primary focus to behavioral studies of more primitive laboratory animals.

A Re-evaluation of Introspection

A century ago, James commented on the status of the psychology of his day as follows.

It must be frankly confessed that in no fundamental sense do we know where our successive fields of consciousness come from, or why they have the precise inner constitution which they do have. They certainly follow or accompany our brain states, and of course their special forms are determined by our past experiences and education. But, if we ask just how the brain conditions them, we have not the remotest inkling of an answer to give; and, if we ask just how the education moulds the brain, we can speak but in the most abstract, general, and conjectural terms.[25]

Since that time, especially during the past few decades, neuroscientists have made major advances in discovering ways in which the brain influences mental processes, but they remain in the dark as to the origins of states of consciousness and the nature of their precise inner constitution. Moreover, it is important not to overlook the fact that neuroscience has made such progress in part by relying on subjects' firsthand accounts of their own mental states. Study of the brain by itself, without reliance on subjective accounts of mental phenomena, can reveal very little about the relationship between the brain and mind.

Thus, insofar as behaviorist and neuroscientific models rely on firsthand accounts of experience, they continue to depend on introspection. But while marvelous advances in technology and methodologies have been made for studying the brain, no advances have been made in refining individuals' introspective abilities. As long as this lack of parity continues, the fallibility of firsthand observations and accounts of subjective experiences can only limit their contribution to these other approaches to understanding the mind and its relation to the brain and behavior. If we could substitute for introspection some technology that could actually detect consciousness and other subjective mental phenomena, this could be a real option. But we now lack such technology, and scientific materialists' *faith* in future neuroscientific breakthroughs is no substitute for our present *knowledge* that we do have some introspective access to subjective mental states. While untrained introspection is like unrefined gold ore, neuroscientific understanding of the origins, nature, and causal efficacy of consciousness is like an undated check. Surely it makes more sense to refine a mode of inquiry that has some direct access to mental phenomena rather than to rely exclusively on another mode of inquiry that has, by itself, provided no evidence even for the existence of subjective mental phenomena.

Despite the claims of scientific materialists such as Lyons, on the basis of our own experience it is apparent that we are aware—or at least have the ability to be aware—of at least some of our own mental states, such as whether our minds are agitated or calm, angry or serene, frustrated or satisfied, alert or dull. We can detect whether thoughts are present or absent. We can passively attend to mental imagery or create it intentionally. We can ascertain whether we desire something and whether we intend to try to fulfill this desire. And we can sense whether we believe, disbelieve, or doubt a proposition. The real question is not *whether* we have such introspective abilities but how that faculty operates; and a major question for scientific research is whether that faculty can be refined so that it can be used to probe mental phenomena more deeply, clearly, and reliably.

A major impact of scientific materialism upon the study of the mind is that it alienates us from our firsthand experience of our own minds, which it equates with "common sense" or "folk psychology." This dismissal of subjective experience is based on the premise that it is often, if not always, misleading. Scientific materialists tell us we should rather rely solely on scientists' observations of other people's brains and behavior, as if they had no firsthand experiences of their own minds. Scientists have thus been appointed a role comparable to the priests of medieval Christendom, who were deemed the necessary intermediaries between the general public and ultimate reality. Ultimate reality for Christianity is God, and for scientific materialism it is matter, which in this case is the matter constituting the brain.

Mental Perception

Many scientific materialists seem to have so ignored their own firsthand experience of the mind that they fail to recognize that personal experience is not limited to the five physical senses but includes mental perception as well. We are as directly aware of many mental phenomena—such as thoughts, feelings, and mental imagery—as we are of sensory phenomena. A fundamental reason why an experientially and rationally coherent view of introspection eludes such modern, erudite thinkers as Lyons may be that the very idea of mental perception is alien to twentieth-century Western thought. Our common assumption is that perception is confined to the senses, while the mind thinks, feels, desires, intends, remembers, imagines, and so on. But we do not commonly think of the mind *perceiving* any phenomena that are accessible to it alone. The term "mental perception" is not commonly used nowadays. James identifies what happens when a word is lacking:

> We are then prone to suppose that no entity can be there; and so we come to overlook phenomena whose existence would be patent to us all, had we only grown up to hear it familiarly recognized in speech. It is hard to focus our attention on the nameless, and so there results a certain vacuousness in the descriptive parts of most psychologies.[26]

On the basis of our own experience, however, it seems that Augustine was right in asserting the existence of an "inner sense," or mental perception, that is aware of the objects of the outer senses, aware of the acts of seeing and so forth, and aware of mental phenomena such as emotions. We may also infer the existence of mental perception on the basis of the experience of shifting the attention from the center to the periphery of our visual field. With mental perception we are aware that we are seeing, and, without shifting our visual gaze, we can move our attention, or the focus of our mental perception, within the visual field. Or we can shift it from the visual field to sounds, smells, tastes, tactile sensations, and various types of mental phenomena, including diverse forms of mental imagery, feelings, and so on.

Let us place the term "mental perception" within the broader classification of "perception" and use this term to denote a type of experiential awareness with respect to objects of cognition. Such experiential awareness is distinct from conceptual awareness. For example, when I conceptually bring Honolulu to mind while reading travel brochures, the city is not apprehended experientially but by way of generic images, or ideas. In that case, Honolulu—the object of cognition—is indirectly apprehended by means of conceptual cognition; but those mental images based on the travel brochures are directly apprehended with mental perception. Likewise, if I recall my visit to Honolulu last year, the city is again conceptually apprehended by way of generic images, while those images are perceived mentally. However, when I am actually in Honolulu, I experientially apprehend this city with both sensory and mental perception. Moreover, if I dream of being in Honolulu, I experientially apprehend the dream images of Honolulu with mental perception.

While the various types of sensory perception apprehend their objects directly, mental perception apprehends forms, sounds, and so on by the power of the sensory consciousness of them. Moreover, mental perception does not apprehend sensory objects directly; rather, it *recollects* them, for such perception is induced by immediately prior sensory perception.[27]

With this concept of mental perception in mind we are now in a position to ask whether the mind can ever perceive itself. Is it true that in any conscious state we can shift our attention to the state itself? James counters as follows.

> No subjective state, whilst present, is its own object; its object is always something else. There are, it is true, cases in which we appear to be naming our present feeling, and so to be experiencing and observing the same inner fact at a single stroke, as when we say "I feel tired," "I am angry," etc. But these are illusory, and a little attention unmasks the illusion.[28]

According to James, when my attention is focused on the color blue, for instance, I am not observing my *perception* of that color. However, when my interest shifts to my experience of blue, I am in fact *recalling* seeing that color just a moment ago. When I remember seeing that color—whether this happened a year ago or a split second ago—I recall myself observing that color. Thus, when I shift my attention back and forth between attending to the color and remembering seeing the color, it seems as if such a shift is comparable to shifting my attention from the objects at the center of consciousness to those at the periphery. However, according to James, the attention is instead shifted from the perceived object to a remembered event. This account maintains the distinction between the observation and the observed object.

The same may be true of our mental awareness of purely mental, as opposed to sensory, processes. Although mental consciousness can apprehend a wide range of objective phenomena, it seems that it cannot take itself as an object, for as long as the mind is operating within the framework of subject/object dualities, it is impossible for the cognizing agent, the cognized object, and the act of cognition to be identical. However, when we remember experiencing a certain mental event, don't we have the ability to recall both the perceived event and our own perception of that event? It seems that the mind recalls the object and subject in an interrelated fashion, even without being conscious of its own presence as the perceiver during the original experience. Thus, even our consciousness of mental feelings of joy, sorrow, frustration, and so on, which seem to be immediately present, may in fact be very short-term memories of mental events occurring just prior to our awareness of them.

There are other mental phenomena, including mental images of visual forms, sounds, smells, tastes, and tactile sensations, that never cognize their own objects and are perceived only mentally. For instance, a mental image of a triangle is perceived as an object of mental perception, but that image itself does not cognize anything. While intentional mental processes—that is, those that apprehend their own objects—may be perceived only retrospectively, these nonintentional mental phenomena and the mental perception of them arise simultaneously. Thus, we are only retrospectively aware of intentional states of consciousness, but we are immediately aware of nonintentional mental phenomena.

The Fallibility of Introspection

Descartes's views on introspection certainly had a major influence on its traditional philosophical use in the West, and when the flaws of his theory became evident during the heyday of introspectionism, that supported the rejection of this mode of observation altogether. According to Descartes, judgments based on introspection can err only insofar as we judge that the events "inside" our minds resemble things "outside" our minds or are modeled on them. Insofar as our introspective judgments remain "internal," he believed, there can hardly be any room for error. Moreover, he hypothesized that everything that is grasped distinctly and clearly must necessarily be true; "no contrary reasons can be adduced to make me doubt my conclusion," he claimed, "and thus my knowledge is true and certain."[29] Although he admitted to having erred in the past, he retrospectively concluded that in those cases he was not seeing clearly and distinctly and had accepted reasons that were not valid.

Descartes's absolutist position collapses in the light of experience, for it is evident that we can misidentify the nature of our own mental events, without reference to any external phenomena. For example, John may believe that he feels only the most selfless love for Mary, whereas in fact his feelings for her are aimed at his own self-gratification. Or Peter may believe that he is motivated to become a social activist out of pure altruism, whereas in reality he is motivated largely by hatred and resentment. Likewise, we may misidentify our own mental imagery, just as we can misidentify the contents of our sensory perceptions. Descartes claims that even in one's sleep, all that presents itself with evidence to the mind is absolutely true.[30] But within a dream situation, we can mistake one person for another and make any number of other mistakes. The only way to salvage Descartes's position may be to retroactively conclude that all such cases of misidentification were actually lacking in "clear and distinct" perception. But this leaves us with nothing more than a tautology: I know that what I am perceiving is real and true because my perception of it is clear and distinct; and my perception is clear and distinct because what I am perceiving is real and true.

In complete opposition to Descartes, James asserts that introspection is a difficult and *fallible* method of examining mental states; but he adds that

> the difficulty is simply that of all observation of whatever kind. . . . The only safeguard is in the final consensus of our farther knowledge about the thing in question, later views correcting earlier ones, until at last the harmony of a consistent system is reached.[31]

This is precisely how scientific observations have always been validated. The fact that introspective observations are private and may be specific to the individual should not detract from their value, he maintains, for in fact no point of view is absolutely public and universal. Private and incommunicable perceptions are inescapable.[32]

James's evaluation of introspection within the broader framework of perception has been rejected by some scientific materialists on the grounds that unaided, human sensory perception is fundamentally unreliable and on these grounds alone introspection has no place in scientific research. However, any instrument of detection has finite sensitivity, specificity, reliability, and precision, and its usefulness is determined in terms of its purposes. Thus, for many of the purposes to which the human senses are applied, they are perfectly adequate and reliable, while in other circumstances they are not. For example, from a neurophysiological perspective, central control of sensory receptors and central sensory relays modifies incoming sensory signals before they reach levels of perceptual experience. These modifications are not random effects but systematically relate to past experiences, expectations, and purposes of the perceiver. These powerful and ubiquitous mechanisms are built into our nervous systems in accordance with genetic, nutritional, and experiential contexts in which we grow up. This condition is equally true of a scientist making observations with a microscope, a surveyor observing a landscape, and a psychological subject introspectively observing mental events.

As a result of such influences, two persons may give contradictory testimony to witnessing "the same event," as it would ostensibly be seen from a purely objective standpoint. Many neuroscientists conclude that one or both of them is either lying or is unable to go from the "raw" (that is, ideally objective) percept to testimony without subjective, psychological distortion. However, others suggest that one or both of them are perceiving the event with sufficient modulation prior to the assembly of their perceptions that, in effect, the two are observing two distinctively different events. This problem pertains equally to extraspective, sensory perception and introspective mental perception, and it is one more instance of *underdetermination*, which is prevalent in scientific research.

Despite its evident fallibility, might there be some facet of clear and distinct perception that is infallible, as Descartes proposed? On the basis of our own experience, all mental images are evident to the perceptions to which they appear; so all perceptions may be said to be valid simply with respect to those appearances. Even in the case of a mistaken cognition, such as mistaking a coiled rope for a snake, there is an *appearance* of the non-existent object; and that appearance of a snake does exist. Thus, a mistaken cognition errs only in terms of how it apprehends or conceives of its object. But all cognitions—including sensory and mental perception, as well as all types of conceptual cognitions—may be said to be valid with reference to the representations that directly appear to them. If so, all mental perceptions are valid with respect to the *appearances* of mental phenomena; but they, like any other perception, may be mistaken in the identification of those phenomena.[33]

This hypothesis differs from the Cartesian view in its assertion that if one attends with mental perception to a mental representation itself, there is room for error in the manner in which one apprehends that object. The distinction may be drawn here between *perceiving* and *identifying as*. According to the hypothesis presented here, mental perception is always valid with respect to the *mere appearance* of mental representations; but it may be mistaken in the manner in which it identifies them. Although the mental perception of appearances is valid, error may creep in as soon as one identifies those appearances as being one thing and not another. In the case of dreaming, there is commonly the mistake of apprehending dream phenomena as existing independently of the dreaming mind. For example, in a dream I may mentally perceive an image of a unicorn. That perception is valid. But if I apprehend it as a real unicorn existing independently of my perception of it, that cognition is mistaken. On the other hand, if I identify the dream state for what it is, and I identify it as a unicorn in a dream, that cognition is valid.

In our day-to-day experience it is evident that we almost always perceive phenomena as familiar things and events that are intelligible within our conceptual frameworks. Thus, in virtually all our perceptions—both extraspective and introspective—there is the possibility of error. Insofar as a perception is theory laden, it is in principle fallible; but it is also our theories that allow us to perceive many phenomena that would otherwise remain hidden from view. A well-trained molecular biologist will see many things with a microscope that are not seen with the untrained eye, and a person who is well versed in a sophisticated theory of mental phenomena may be able to observe introspectively many things that would otherwise be hidden. In all kinds of perception, as James asserts, it is crucial to learn how to observe *relationships*, as well as discrete entities; and in the case of introspection this may be essential to the observation of such intentional conscious states as desires, beliefs, intentions, and other mental phenomena that are commonly considered to be unobservable.

Theoretical training can be a great aid in cultivating one's introspective faculties, but that alone does not suffice. As an analogy, no matter how educated a cell biologist may be, if the microscope used in research is mounted on an unstable platform so that it frequently jiggles; if its optical system has poor resolution; and if the subject under examination is poorly illuminated, it will be impossible to collect reliable empirical data. Likewise, if the attention of a person practicing introspection is frequently agitated, and if there is little clarity or precision in one's introspective observations, then the reliability of this mode of inquiry is undermined. Let us turn now to the question of how to refine the attention so that the mind may be a more effective instrument in observing that range of phenomena that is accessible to it alone.

EXPLORING THE MIND

William James on Sustained, Voluntary Attention

Scientific materialists commonly regard introspection as being inadequate to the task of providing reliable data that can generate anything approximating scientific consensus. Scientific research, they point out, requires systematic methods of observation capable of revealing phenomena and dynamics that are often inaccessible to unaided observation. Indeed, the development and employment of such empirical tools of research have been instrumental in enabling modern science to leave behind false assumptions based on common sense as well as antiquated dogmas.

Although the presence of mental phenomena can be inferred on the basis of third-person evidence, they can be observed by one means only: mental perception. But this mode of observation, like all others, is fallible. Science has sought to avoid the limitations and unreliability of human perception by developing mechanical instruments to enhance and supplant our sensory awareness of the physical world. But during the course of this amazing progress in technology, no comparable scientific advances have been made in refining our own human abilities of attention and perception. The lopsided nature of this progress is easily overlooked as long as we attend solely to external, physical phenomena; but it becomes glaringly evident as soon as we try to focus inward on our own mental processes. For extraspective scientific research, scientists must be confident that their tools of observation and analysis are reliable and appropriate for the object under investigation. Specifically, they must be assured that their empirical data actually originate from the object under study and are not simply artifacts of their data-gathering apparatus. Similarly, if introspection is to become a reliable means of investigating the mind, researchers must be able to observe mental phenomena with a high degree of attentional stability and vividness. In this regard, the importance of sustained, voluntary attention cannot be overemphasized.

William James placed the issue of sustained, voluntary attention in the broader context of education and human life as a whole. The first and most important task at the outset of an education, he argued, is to overcome gradually the inattentive dispersion of the attention. In education, "the power of voluntarily attending is the point of the whole procedure. Just as a balance turns on its knife-edges, so upon it our moral destiny turns."[1] In his classic work *The Principles of Psychology*, he elaborates further on this theme:

> the faculty of voluntarily bringing back a wandering attention, over and over again, is the very root of judgment, character, and will.... An education which should improve this faculty would be *the* education *par excellence*. But it is easier to define this ideal than to give practical directions for bringing it about.[2]

James saw a particularly deep relationship between the attention and the will. In another expression of his attentional reality principle, he commented that "each of us literally chooses, by his ways of attending to things, what sort of a universe he shall appear to himself to inhabit."[3] He regarded as the essential achievement of the will the ability to attend to a difficult object and hold it fast before the mind. It is noteworthy that over the past forty years modern psychology, under the influence of the closure principle, has largely ignored the topic of the will.[4] For James, the implications of this interrelation can hardly be exaggerated.

> When we reflect that the turnings of our attention form the nucleus of our inner self; when we see ... that volition is nothing but attention; when we believe that our autonomy in the midst of nature depends on our not being pure effect, but a cause, ... we must admit that the question whether attention involve such a principle of spiritual activity or not is metaphysical as well as psychological, and is well worthy of all the pains we can bestow on its solution. It is in fact the pivotal question of metaphysics, the very hinge on which our picture of the world shall swing from materialism, fatalism, monism, towards spiritualism, freedom, pluralism,—or else the other way.[5]

James also linked the attention, given its intimate relation with volition, to morality. A moral act, in its simplest and most elementary form "*consists in the effort of attention by which we hold fast to an idea* which but for that effort of attention would be driven out of the mind by the other psychological tendencies that are there."[6] Thus, in James's view, sustained, voluntary attention not only is crucial in the practice of introspection, which he regarded as the foremost means of examining mental phenomena, but also exerts a powerful influence on the types of reality we experience and on our moral conduct.

While James attributed a tremendous importance to the attention, he was fundamentally pessimistic as to the possibility of improving it. In *The Varieties of Religious Experience* he does comment briefly and sympathetically on the Hindu practice of *samādhi* and the Buddhist experience of *dhyāna*, both of which entail sustained, voluntary attention.[7] However, in his time the paucity of authoritative, lucid works in European languages on Hindu and Buddhist contemplative practice made it very difficult for Western scholars to appreciate the real significance of these states. Thus, James makes no reference to these in his psychological writings on attention.

On the basis of the best scientific data of his day, James concluded that it is impossible to sustain the attention voluntarily for more than a few seconds at a time. Moreover, he declared that it is possible only in pathological cases for someone to attend continuously to an object that does not change.[8] Genuine sustained, voluntary attention, on the contrary, is brought about by repeatedly focusing the attention on a *developing* topic. In this way, attention may be sustained, under favorable conditions, for hours on end. He emphasized, however, that such a topic is actually a succession of mutually related objects that form one topic to which the attention is directed. James noted in passing that geniuses are commonly thought to excel in this ability; but he assumed that it is their genius making them attentive, not their exceptional faculty of attention making geniuses of them.

James noted that some people seem to be naturally scatterbrained, while others can apparently follow a train of connected thoughts without distraction. He adds that

[t]he possession of such a steady faculty of attention is unquestionably a great boon. Those who have it can work more rapidly, and with less nervous wear and tear. I am inclined to think that no one who is without it naturally can by any amount of drill or discipline attain it in a very high degree. Its amount is probably a fixed characteristic of the individual.[9]

While psychologists of James's day studied attention quite closely, and it continues to be a major topic of psychological research now, relatively little scientific research has been conducted to test the hypothesis that the stability and vividness of attention is fixed and untrainable. If any of James's claims concerning the significance of sustained, voluntary attention are valid, this oversight appears to be all the more significant.

According to modern neuroscientific studies of attention, the effects of the highest levels of attention on outwardly manifested performance of simple sensory tasks are not typically sustained for more than one to three seconds. However, while focused attention may be enhanced for only that brief duration without additional stimuli or other external assistance, there is evidence that relatively high levels of attention may be sustained for many tens of minutes.[10] Because of the paucity of collaboration between scientists and experienced contemplatives, it is not clear whether the kinds of attention cultivated by Hindu and Buddhist contemplatives are that "highest level" that, according to scientific research, can be maintained for only a few seconds or whether they are the type that can commonly be sustained for much longer periods. What can be said is that the experiments that indicate that the transient, high level of focused attention lasts between one and three seconds[11] corroborate James's conclusion more than a century ago.

While psychologists have apparently assumed that normal attention cannot be enhanced through training, the recent proliferation of a wide array of attentional disorders generically labeled "attention deficit disorders" among school children, adolescents, and adults has made it all too obvious that one's attentional abilities can deteriorate. As a society, it seems that the modern West is not only failing to make progress with regard to sustained, voluntary attention, we are actually losing ground. Treatment for attention deficit disorders (ADD) generally ranges from medicinal to behavioral to environmental interventions. Predictably, given the domination of our society by scientific materialism, very little has been done in terms of devising cognitive methods for restoring one's attentional abilities. Rather, the treatment most often used for people with ADD is medication therapy, which has frequently proven effective for short-term increases in attentional stability but offers no long-term benefits for increasing patients' attentional capacities.[12] Furthermore, literature on the effects of such drugs does not indicate that they help strengthen the will, nor does it distinguish whether improvement is the result of the medication (1) helping people to realize when they are drifting off the topic and to bring their attention back or (2) helping them stay on the topic in the first place.

The drug most commonly prescribed for children with attention deficit disorders is Ritalin, which is also prescribed for adults suffering from narcolepsy. Possible short-term side effects of this medication include nausea, anorexia, headache, blood pressure and pulse changes (either up or down), palpitations, nervousness and insomnia, drowsiness, cardiac arrhythmias, and abdominal pain. Possible long-term side effects include weight loss, growth retardation, insomnia, and addiction. In 1998, the number of prescriptions for Ritalin made for children in the United States under the age of twelve was 2,400,000, and this number is quickly rising. In prescribing this drug, physicians are warned that Ritalin is contraindicated for patients with psychological problems such as depression, psychosis, or chronic fatigue. If they do have other psychological problems, other drugs, with their own daunting sets of possible side effects, are prescribed. In the United States 20 percent of children up to the age of twelve are believed to suffer emotional problems, for which they may eventually be given a variety of powerful drugs to suppress their symptoms. Perhaps the most serious long-term effect of taking such medications is that children are encouraged to get in the habit of taking drugs to change the state of their minds, even though the medications themselves may not be addictive.[13]

Over the past century—let us call it the Century of Scientific Materialism—the cognitive sciences have made no progress in developing means of enhancing either the stability or vividness of attention beyond normal levels. In the meantime, attention deficit disorders are becoming more and more endemic, and our primary treatments for them are medications that have no long-lasting benefits but do have serious side effects. Psychologist James Hans comments as follows.

> That our life passes with more or less attention to its passing is ... obvious. That the richness of life is a function of our full attentiveness to what is goes without saying as well.... If attention is truly all we have and all we are ... then we need to reconsider our relationship to the most fundamental feature of our lives.[14]

In reconsidering our relation to attention, given the paucity of our own scientific resources, it is only reasonable to look beyond our own contemporary society to the wisdom of earlier eras of our own culture and to other cultures that have not been encumbered by the dominant ideology that so constrains modern scientific and medical research.

Augustinian Contemplative Inquiry into Consciousness

Augustine is only one of many contemplatives from around the world to have practiced means of refining mental perception so that it can be used more effectively in exploring the nature of consciousness and other mental phenomena; and it seems he did so by developing sustained attention with consciousness itself as his object. While the contemplative training he advocates culminates in genuine contemplation, in which the soul is said to transcend itself in the experience of God, the proximate preparation to that stage consists of two processes, *recollection* and *introversion*. Recollection entails withdrawing the attention from all sensory and conceptual phenomena; and the subsequent training in introversion consists of introspectively settling the mind in its own fundamental nature.[15] In a similar vein, the early Christian desert fathers spoke of a state called *apatheia*, a kind of dispassionate serenity in which the intellect rises above distraction and attains its natural state. This is not something the mind *does* but rather something it *is*, hence it is called a state (*katastasis*).

The feebleness of the attention as an obstacle in the pursuit of contemplative insight has been widely recognized since the early centuries of Christianity. One of the clearest descriptions of the ordinary—which is to say, attentionally dysfunctional—mind is the immensely popular and seminal fifth-century work *The Conferences of Cassian*. Cassian's Abba Moses says of the undisciplined mind that it "flutters hither and thither, according to the whim of the passing moment and follows whatever immediate and external impression is presented to it," while thoughts "career about the soul" like bubbling, effervescent "boiling water."[16] The same observation was made almost a millennium later by the German Christian contemplative Meister Eckhart (1206?–1327?), who spoke of the "storm of inward thoughts" that had to be calmed in the course of one's contemplative training.[17] Like Augustine, Eckhart taught that in order to fathom the nature of one's own being, "a man must collect all his powers as if into a corner of his soul . . . hiding away from all images and forms."[18] By calming attentional excitation and stilling the mind utterly, Eckhart claimed, one experiences a state of "rapture" (Gezücken), in which the contemplative finds himself "in a state in which there are no images and no desires in him, and he will therefore stand without activity, internal or external."[19]

William James rejects the notion that we can introspectively "capture our consciousness life-like, as a pure spiritual activity, neglecting almost completely the materials which consciousness illuminates at any given moment."[20] However, Augustine and later Christian contemplatives maintained that while the nature of the mind cannot be discovered by observing external phenomena, it can be discovered by withdrawing the mind from all appearances that have been added onto it. Once the mind dispenses with all such phenomena, including speculative ideas about itself, and simply encounters its own inward presence, then that which remains is the mind itself.[21]

Augustine raises a number of questions concerning the manner in which the mind observes itself: Since the mind is never without itself, why does it not always observe itself? Does one part of the mind observe another part? In the process of introspection does the mind serve as both its subject and its object? To all such questions he responds that such notions are artificial, conceptual constructions "and that the mind is not such is absolutely certain to the few minds that can be consulted for the truth about this matter."[22]

The rise of modern science coincided with the decline of contemplative practice in Western Christianity. The tradition of the Christian Church during the time of Augustine was that contemplation is the central objective of the spiritual life and thus is open to everyone as something to be aspired to and practiced. However, by the eighteenth and nineteenth centuries, even Roman Catholicism came to look upon contemplation as something peculiar, often associated with visions, revelations, raptures, stigmatization, levitation, and other bizarre psychophysical phenomena. The Church warned that contemplation was at best to be admired from a safe distance, for its actual practice was dangerous and full of pitfalls.[23] From the nineteenth century onward, mysticism has been characterized as abounding in revelations and visions, mostly by women; and it has been linked by Western society with extravagance, fanaticism, and delusion. Thus, contemplation has commonly been regarded by the Roman Catholic Church as something that is out of reach for all but a few specially called and favored souls. In contrast to early forms of Christianity, the Protestant movement, on the whole, has the distinction of never having endorsed any contemplative discipline. Indeed, for this school of Christianity, which grew up together with scientific materialism, it looks as if scientific inquiry and knowledge have taken the place of contemplative inquiry and knowledge.

The Buddhist Cultivation of Sustained, Voluntary Attention

Many contemplative traditions from around the world and throughout history—including those of Judaism, Christianity, Hinduism, and Buddhism—have proposed that a primary means for gaining experiential knowledge of ultimate reality is through acquiring firsthand insight into the nature of consciousness.[24] William James, whose sympathetic studies of mysticism were confined largely to the Christian tradition, lists four common characteristics of contemplative experience: ineffability, a noetic sense of being in contact with ultimate reality, transience, and a passive stilling of the mind.[25] Based on his research of mystics' accounts of their own experiences, he concluded that such contemplative states *cannot* last. Within the Christian tradition, the insistence on the fleeting nature of mystical union appears to originate with Augustine[26] and is reflected in the writings of Meister Eckhart, who emphasized that the state of rapture is invariably transient, with even its residual effects lasting no longer than three days.[27]

Hindu and Buddhist contemplatives, on the contrary, claim that even the most sublime contemplative states may be sustained for hours or even days on end and their residual effects may be permanent. The standard Christian interpretation of the fleeting nature of mystical union is that the very nature of the human soul is such that it cannot sustain its encounter with the divine for more than a brief time. Buddhist contemplatives, on the other hand, claim that if one cannot sustain one's experiential realization of ultimate reality, this is due to an inadequate prior cultivation of sustained, voluntary attention (samādhi).

Vision-Induced, Sustained Attention

Unlike modern science, the contemplative traditions of the world have long devoted themselves to the challenge of developing means of refining mental perception with various methods of training the attention, and they have concluded that the faculty of attention is not a fixed characteristic of the individual, as James assumed. One array of such techniques that is most closely connected to our sensory experience of the physical world has long been practiced in Theravāda Buddhism, which remains a living tradition to this day in southeast Asia. Buddhaghosa, an Indian contemporary of Augustine and the most authoritative commentator in the Theravāda Buddhist tradition, gives an elaborate account of techniques for developing sustained attention using emblems of various elements of sensory experience. Specifically, he lists ten types of emblems corresponding to the four elements of earth (solidity), water (fluidity), fire (heat), and air (motility); the four primary colors of blue, yellow, red, and white; and finally light and space.

To describe briefly one example of such practice, in the case of focusing on an earth emblem, one first fixes one's gaze on a disc prepared of clay representing the entirety of the element of earth, or solidity. One repeatedly gazes at this device until an afterimage of it appears in the mind as clearly when the eyes are shut as when they are open. This mental image, the *sign* of the earth element, becomes the chief meditative object during the preliminary training in sustained attention. When one first crosses the threshold into meditative stabilization (dhyāna), there arises to the mind's eye a more refined sign of the earth element. This image, the *counterpart sign*, is an appearance that arises purely from mental perception. It has no color, no appearance of solidity, and none of the blemishes of the original earth emblem that were evident in the earlier mental image. This counterpart sign is regarded as a mental representation of the primal quality of the element of earth.[28]

In this Theravāda account, the development of sustained attention is closely linked to three kinds of signs that are the objects of one's attention. The first of these is the sign for preliminary practice, which in the case of the earth emblem is the actual physical symbol of earth used for this practice. The second is the acquired sign, which in the case of the earth emblem is the afterimage that appears as a precise copy of the first sign, with all its specific limitations, such as its molded form, color, and shape. The third is the counterpart sign, which is a subtle, emblematic representation of the whole quality of the element it symbolizes.

Theravāda Buddhist contemplatives claim that physical reality may be mentally altered by the contemplative manipulation of the counterpart signs. The role of meditative stabilization in the discipline of contemplation may be likened to the role of mathematics in the physical sciences. Without knowledge of mathematics and the ability to apply this knowledge in the study of the laws of nature, modern physical science would hardly have progressed as it has. Mathematics is indispensable not only for scientific understanding of the physical world but also for developing the necessary technology to further our knowledge and control of nature. Similarly, meditative stabilization is said to be indispensable for gaining contemplative insight into the nature of physical and mental phenomena; and it is said to allow for the development of various types of extrasensory perception and paranormal abilities that can be used in knowing and controlling nature.

Buddhaghosa explains in detail how the mind is exercised in the use of counterpart signs in order to develop extrasensory perception and paranormal abilities.[29] To take one example, if one wishes to transform a liquid into a solid, one focuses on the counterpart sign of the earth emblem. Then, on emerging from the state of meditative stabilization, one focuses the attention on a body of liquid, such as a lake, and resolves, "Let there be earth"; and it becomes solid, so that one can walk upon it freely.[30] This contemplative tradition claims that this exertion of the mind's power over matter can be either private or public, as the contemplative wishes. Thus, abilities such as walking on water and multiplying physical objects are seen not as acts of supernatural intervention but as rational, lawful manipulations of matter by the mind. The fundamental hypothesis is that consciousness is an integral element of the natural world and that it holds extraordinary capacities that run completely counter to commonsense experience.

The Theravāda tradition asserts that after the counterpart sign appears and vanishes, one experiences the primal state of the mind from which thoughts originate. This state of awareness is said to be "process-free," in contrast to the "active mind," and as it is free from all sense impressions, it shines in its own radiance, which is otherwise obscured because of external influences.[31]

According to this ancient contemplative tradition, it is possible to train the mind so that the attention can be uninterruptedly sustained for hours on end. Such concentrative ability is said to be crucial for fathoming the nature of consciousness and tapping its hidden potentials. These claims of Theravāda Buddhist contemplatives obviously appear incredible in light of our commonsense assumptions about the mind. Moreover, our indoctrination into scientific materialism tells us that such claims must be false as a matter of principle, without our ever putting those training techniques to the test. Nevertheless, the simple fact is that Western scientists have never conducted the kind of research on developing sustained attention that was done in ancient India and continues to be pursued in southeast Asia today. It is experience alone—not the metaphysical assumptions of Buddhism or scientific materialism—that can determine whether the claims of this contemplative tradition are valid.

Imagination-Induced Sustained Attention

While Buddhaghosa relied heavily on Singhalese accounts of Buddhist contemplative practice, two of his Indian contemporaries, Vasubandhu and Asaṅga, belonged to another Buddhist tradition whose records were preserved in Sanskrit. These two contemplatives, who were brothers, are among the most authoritative proponents of the school of Mahāyāna Buddhism, which remains today a living tradition among Tibetans and other Asian societies. Like Buddhaghosa, they assert that, contrary to James's belief, the healthy mind can in fact attend continuously to an object that does not change. However, while one focuses the attention on an unchanging object, there is the possibility of dementia setting in if one allows the potency of attentional vividness to wane. The result of such faulty practice is that one enters a kind of trance, or mental stupor, in which one's intelligence degenerates. The way to avert this danger is by taking on the difficult challenge of enhancing one's attentional vividness without sacrificing attentional stability.

The Mahāyāna Buddhist contemplative tradition uses a wide array of objects for the cultivation of sustained attention, but it especially emphasizes the practice of visualization.[32] Here a clear distinction must be made between the *physical support* for the meditative object and the meditative object

itself. Any kind of physical object, commonly one with religious significance, may be used in the preliminary stages of this practice so that one
becomes thoroughly familiar with its characteristics. But during the actual
training in sustained attention, one visualizes a mental image of that object.
Unlike the previously described technique of attending to a visually induced
afterimage, this method entails mentally creating and sustaining an image
of a physical object, based either on seeing it or on hearing of its characteristics. Rather than viewing a two-dimensional mental image, one imagines the object three-dimensionally, bringing to mind its qualities on all its
surfaces. It is commonly asserted in this tradition that the deepest states of
samādhi, or single-pointed concentration, can be attained only when the
attention is directed upon a mental object, for samādhi is accomplished
with mental, not sensory consciousness.

In this form of training, two qualities must be cultivated: attentional
stability and vividness. To understand these two qualities in terms of Buddhist psychology, one must note that Buddhist contemplatives commonly
assert that the continuum of awareness is composed of successive moments,
or pulses, of cognition having finite duration. Vasubandhu asserts that the
duration of a single moment of awareness is between one and two milli-
seconds.[33] For the meditatively untrained mind, however, due to its extreme
brevity, no single moment of awareness has the capacity of ascertaining
anything. Moreover, in a continuum of perception, many moments of
awareness often consist of nonascertaining cognition; that is, objects *appear*
to this inattentive awareness but they are not *ascertained*, so one cannot
recall them later on.

In terms of this theory, the degree of attentional stability increases in
relation to the proportion of ascertaining moments of cognition of the intentional object. That is, as stability increases, fewer and fewer moments of
ascertaining consciousness are focused on any other object, making for a
homogeneity of moments of ascertaining perception. The degree of attentional vividness corresponds to the ratio of moments of ascertaining to nonascertaining cognition: the higher the frequency of ascertaining perception,
the greater the vividness. Thus, the achievement of meditative stabilization
entails an exceptionally high density of homogenous moments of ascertaining consciousness. In the contemplative cultivation of sustained attention it
is not enough that one's attention is stable and vivid; rather, one must
ascertain the meditative object. Otherwise, the full potency of attentional
vividness cannot arise, and one's samādhi remains impaired.

In order to develop attentional stability and vividness, two mental faculties must be cultivated: mindfulness and introspection. The task of mindfulness is to attend without distraction to a familiar object of attention,
while the function of introspection is to monitor the attending awareness.
Mindfulness is the most important factor for developing introspection and
is the principal means of accomplishing meditative stabilization. When the
power of mindfulness has fully emerged, the attention no longer strays from
its object. At that time, if one does not continue striving to enhance the

power of attentional vividness, one may fall into a complacent, pseudo-meditative trance, which may result in dementia.

The two chief obstacles to the cultivation of attentional stability and vividness are excitation and laxity, respectively, and it is the task of introspection to detect the occurrence of these mental processes as soon as they arise. Thus, introspection is often likened to a sentry who stands guard against these two hindrances. Excitation is defined as an agitated mental state, driven by desire for pleasurable stimuli, which acts as an obstacle to meditative stabilization. Laxity, on the other hand, is said to arise from lethargy and occurs when the attention is slack and one does not apprehend the meditative object with vividness or forcefulness. Excitation is easy to recognize, but since laxity is difficult to identify, under its influence one may easily overestimate the quality of one's attention.

The purpose of mindfulness is first to prevent the attention from being distracted from the meditative object. When *subtle excitation* arises, it may seem as if one's attention is continuously focused on the meditative object even while the mind is peripherally distracted to other objects; whereas in the case of *coarse excitation* the meditative object is forgotten entirely. However, according to Buddhist psychology, a single moment of consciousness cannot attend simultaneously to two or more dissimilar objects. Thus, subtle excitation must entail successive moments of cognition of the meditative object briefly interrupted by cognitions of other objects. As these moments of cognition are experientially blurred together, one may have the mistaken impression that, despite these distractions, there is an unbroken continuity of awareness of the meditative object. In the case of coarse excitation, the continuity of attention focused on the meditative object is interrupted so long that one notes that the meditative object has been forgotten altogether.

As a result of diligently counteracting even the most subtle laxity and excitation as soon as they occur, eventually effortless, sustained attention is said to arise due to the power of habituation. At this point, only an initial impulse of will and effort is needed at the beginning of each meditation session; thereafter, uninterrupted, sustained attention occurs effortlessly. Now it is actually a hindrance to engage the will or to exert effort. It is time to let the natural balance of the mind maintain itself without interference. In this state of meditative stabilization, because of the extraordinarily high degree of stability and vividness of the attention, the imagined visual object appears before the mind's eye with almost the brilliance of a visually perceived object.

When meditative stabilization is finally achieved, it is said that the entire continuum of one's attention is focused single-pointedly, nonconceptually, and internally in the very quiescence of the mind; and the attention is withdrawn fully from the physical senses. At that point, if occasional thoughts do arise, even about the meditative object, Asaṅga advises the trainee not to follow after them. Thus, one now disengages not only from extraneous thoughts and so forth but even from the meditative object. For the first time in this training, one does not attempt to fix the attention upon a familiar object. One's consciousness is now left in an absence of appearances, an experience that Asaṅga says is subtle and difficult to realize.

Upon achieving this meditative state, both Asaṅga and Vasubandhu assert, the mind disengages from the representations of sensory objects, and only the aspects of the sheer awareness, luminosity, and vivid joy of the mind appear. Thus, these contemplatives, in concert with Jewish, Christian, and Hindu contemplatives, present us with the truly astonishing hypothesis that joy arises from the very nature of consciousness once it is free of the afflictions of laxity and excitation and is disengaged from all sensory and mental appearances. In this state, any thoughts that arise neither are sustained nor do they proliferate; rather they vanish of their own accord, like bubbles emerging from water. One has no sense of one's own body, and it seems as if one's mind has become indivisible with space. This state is characterized as one of joy, luminosity, and nonconceptuality.

Asaṅga asserts that with the achievement of meditative stabilization one cuts through one's culturally and personally acquired conceptual conditioning, including the sense of one's own gender, and experientially fathoms the nature of the mind. He does not present this as a uniquely Buddhist comprehension of the mind, nor does he regard it as a comprehension of a uniquely Buddhist mind. Rather, in this state one gains a transcultural and transpersonal realization of the nature of consciousness. In the state of meditative stabilization, the mind is no longer consciously engaged with human thought, mental imagery, or language, and it is disengaged from the human senses. Thus, this training is presented as a means for experientially ascertaining the nature of consciousness itself, which is common to people of different cultures and times and to human and nonhuman sentient beings.

This assertion need not be interpreted as contradicting the hypothesis that consciousness cannot apprehend itself. That premise denies that a single consciousness can have itself as its own object. During the development of meditative stabilization, introspection has the function of monitoring the meditator's consciousness, particularly regarding the occurrence of the mental processes of laxity and excitation. Such metacognition is a form of *recollective* awareness that cognizes previous moments of consciousness. Likewise, once meditative stabilization is accomplished and one's meditative object dissolves, in this absence of appearances the continuum of one's attention may attend to *previous* moments of consciousness. Because of the homogeneity of this mental continuum, the experiential effect would be that of the mind apprehending itself.

Mahāyāna Buddhism, like the Theravāda tradition, declares that various types of extrasensory perception and paranormal abilities can be readily developed once meditative stabilization has been achieved. But in both traditions the chief purpose of developing sustained attention is to acquire insight into the nature of reality; and the nature of the mind is characteristically of principal interest. Vasubandhu comments on the difficulty of this endeavor:

Subtle, unquestionably, are the specific characteristics of the mind and its mental processes. One discerns them only with difficulty even when one is content to consider each of the mental processes as developing in a homogenous series; how much more so when one envisions them in the [psychological] moment in which they all exist. If the differences of the taste of vegetables, tastes that we know through a material organ, are difficult to distinguish, how much more so is this true with non-material phenomena that are perceived through the mental consciousness.[34]

Inverting Awareness

If it is possible to monitor the quality of one's attention while developing sustained attention, and if it is possible to attend to previous moments of consciousness free of appearances after achieving that state, might it also be possible to develop sustained attention with consciousness itself as one's object? Such a technique is commonly practiced in the Great Perfection (Dzogchen) tradition of contemplation first promulgated in Tibet in the eighth century by the Indian Buddhist contemplative Padmasambhava. In a method referred to by the term *maintaining the attention in nonconceptuality*, the mind is withdrawn from the physical senses, as well as all thoughts concerning the past, present, and future. One lets the mind come to rest like a cloudless sky, clear, luminous, and with no intentional object apart from its own presence. Mindfulness and introspection are instrumental in this technique, and one must guard against laxity and excitation as explained earlier. Whenever a thought or any kind of mental imagery arises, one does not follow after it but releases it immediately, leaving one's awareness in the remaining vacuity. When the attention is sustained in that fashion, all mental engagement with other objects is stopped, as if one had fainted or fallen asleep. The crucial difference, however, is that in this meditative state, one is said to ascertain the essential features of consciousness vividly, single-pointedly, and without conceptual mediation. Each time thoughts are detected by means of introspection, they vanish by themselves, leaving only a vacuity in their wake. When the mind is observed free of any conceptual fluctuation, it is seen as an unobscured, clear, and vivid vacuity, without any difference between former and latter states. Tibetan contemplatives call this "the fusion of stillness and dispersion."[35]

Like Augustine, Padmasambhava claims that this method leads to a direct realization of the nature of awareness. What are the defining characteristics of consciousness that is perceived in that way? According to the "Centrist" view (Madhyamaka) advocated by Padmasambhava, all types of consciousness are nonconceptual with respect to their own appearances, so they are said to be imbued with *clarity* regarding those appearances. In this sense, one of the defining characteristics of consciousness is said to be clarity, or luminosity. Because consciousness is experientially *aware* of those appearances, its second defining characteristic is said to be awareness, or cognizance.

In this practice the attention is focused on the *sheer* awareness and the *sheer* clarity of experience, which are the irreducible, defining features of consciousness alone, as opposed to the qualities of other objects of consciousness. Thus, in this technique the object of mindfulness is preceding moments of consciousness; and introspection monitors whether or not the attention is straying from those qualities of the awareness and clarity of experience. As in the previously discussed practice of inducing sustained attention by means of the imagination, this method culminates in the experience of joy, luminosity, and nonconceptuality, which are said to be the natural qualities of the mind at rest.

Tibetan contemplatives, like Augustine, declare that theories about the nature of consciousness and the manner in which introspection functions are indeed artificial, conceptual constructions; for the experience of consciousness when the mind is settled in meditative stabilization is a state in which words and concepts are suspended. Any subsequent theory is nothing more than a conceptual overlay on an experience that is nonconceptual. The point of such theories, however, is to break down conceptual barriers to entering into this experience and to make such realization somewhat intelligible to those who lack it. But no description or explanation can capture this experience in words or concepts or convey the actual nature of the experience to noncontemplatives.

Releasing the Mind

Clarity and awareness are said to be the salient features of consciousness in general, not only of consciousness that is withdrawn from sensory and conceptual stimuli. If this is the case, these qualities should be apprehendable in *all* states of consciousness and not only in the state of meditative stabilization. Moreover, if joy, luminosity, and nonconceptuality are natural qualities of awareness, these should manifest when the mind is left in its natural state and not only when it is strenuously focused on a single object.

It is on this premise that a method known by the term *settling the mind in its natural state* has been widely taught and practiced especially in the Great Perfection tradition of Buddhist contemplation.[36] The distinguishing characteristic of this technique is maintaining the attention without distraction and without conceptual grasping, the latter referring to the mental process of conceptually identifying, or labeling, the objects of the mind. Thus, in this practice, one does not grasp onto the intentional objects of thoughts concerning the past, present, or future, nor does one judge or evaluate thoughts themselves. Now one does not to try to get rid of thoughts but rather observes them nonconceptually. Without identifying the objects of the mind *as anything*, one tries simply to perceive them in their own nature, without identifying them within any conceptual framework. Thus, without conceptually grasping onto the contents of the mind, one perceptually ascertains their clear and cognitive nature; this method, like the previous one, is said to lead to insight into the nature of consciousness itself.

As a result of such practice there arises a nonconceptual sense that nothing can harm the mind, regardless of whether or not ideation has ceased. Whatever kinds of mental imagery occur—be they gentle or violent, subtle or gross, of long or short duration, strong or weak, good or bad—one is to observe their nature, and to avoid any obsessive evaluation of them as being one thing and not another. When this method of maintaining awareness in its natural state is followed precisely, the mind becomes serene, and one does not succumb to disturbing mental processes such as excitation, aggression, anxiety, or resentment.

The afflictions of the mind are naturally calmed when the mind is settled in a state of nongrasping, and the clear and empty nature of awareness is vividly perceived. Whenever thoughts arise, one simply observes them without aversion or approval, and by so doing, thoughts no longer impede the cultivation of sustained attention, nor do they obscure the nature of consciousness. This practice is sometimes elucidated with the analogy of a raven at sea. According to ancient Indian tradition, when a ship went out to sea, a raven was brought along; and when the navigator wanted to know whether he had come near shore, he would release the raven. As in the biblical account of Noah and the ark, if there was no land nearby, the raven would circle around and around, and eventually alight back on the ship. Likewise, in this contemplative practice, one releases the mind so that thoughts flow out freely, without suppressing any of them. As long as thoughts are arising, one observes them without interference, and eventually they disappear, or "alight" back in the nature of awareness from which they originated. With sustained practice, without ever suppressing ideation, the mind becomes still and conceptual dispersion ceases of its own accord. The awareness that is perceived during this practice has no form but is vacuous like space; and yet, like a stainless mirror, it takes on the appearance of all objects that are presented to it.

This practice, unlike all the preceding techniques, allows for a kind of free association of ideas, desires, and emotions. Because one is not intentionally suppressing, evaluating, judging, or directing any thoughts and so on that appear to the mind, and because the attention is maintained within the field of mental phenomena, without being distracted by physical objects, many contents of the unconscious are brought into consciousness. These may include old memories, long-forgotten fears and resentments, repressed desires and fantasies, and so on. As in the dream state, habitual propensities of the mind are catalyzed so that unconscious processes—including those that influence one's behavior, health, and so forth—are made conscious. This method is therefore designed to enhance the depth and scope of one's introspective abilities.

The practice of settling the mind in its natural state also challenges a fundamental assumption of Descartes concerning the nature of one's own personal identity. When this technique is first applied, thoughts tend to vanish as soon as they are detected; but with practice, as one develops a "lighter introspective touch," trains of thought arise, follow their course, and vanish of their own accord. At no point does one have the sense of being the creator, sustainer, or destroyer of thoughts. The sense that "I am thinking that" occurs only when one conceptually grasps onto and identifies with thoughts. This is presumably what led Descartes to conclude that he was a "thinking thing" that doubts, understands, affirms, denies, wills, and feels, "a mind or soul, an understanding or a rational being . . . a real thing, truly existent."[37] Descartes expresses his absolute certainty that the I, or mind, is a real thinking substance, but this assumption is seriously challenged when thoughts are observed arising by themselves without being intentionally created by a thinking agent.

Even when the mind is settled in meditative stabilization without human conceptual constructs, it is not considered by Buddhist contemplatives to be entirely free of all traces of conceptualization. One's inborn sense of a reified self as the observer and the reified sense of the duality between subject and object are still present, even though they may be dormant while in meditation; and when one emerges from this nonconceptual state, the mind may still grasp onto all phenomena, including consciousness itself, as being real, inherently existing entities. To penetrate to the fundamental nature of appearances and their relation to consciousness, it is said that one must go beyond meditative stabilization and engage in training for the cultivation of contemplative insight. Nevertheless, the achievement of meditative stabilization is taught as a crucial prerequisite to gaining conceptually unstructured and unmediated insight into the fundamental nature of reality. Moreover, Padmasambhava warns that without having developed a high degree of attentional stability and vividness, even if one apprehends the nature of awareness, it remains only an object of intellectual understanding, leading merely to philosophical discourse at best and dogmatism at worst.

William James on Pure Experience

The contemplative pursuit of conceptually unstructured awareness may appear to be solely a religious pursuit with little or no relevance to the science of the mind. But in fact William James was keenly interested in a comparable mode of perception that he called pure experience, and he approached this topic by challenging the very existence of consciousness. In his seminal essay "Does Consciousness Exist?" James proposes that consciousness does not exist as a substantial entity, or primal stuff, out of which thoughts are made, and which is utterly distinct from some other primal stuff out of which material objects are made. "Consciousness" conceived as a substantial, subjective substance is "the name of a nonentity, and has no right to a place among first principles."[38] He similarly denies that consciousness is properly conceived as a pure activity, without physical extension, devoid of self-content, but directly self-knowing.

It is important to recognize, however, that James similarly refutes the existence of matter as some primal, objective stuff out of which all physical phenomena arise. Thus, he roundly rejects both consciousness and matter as reified entities.[39] At the same time, he emphatically insists that consciousness does exist as an impersonal function, namely, the function of *knowing*. Concerning the relation between subjects and objects in general, he declares that

> [t]he attributes "subject" and "object," "represented" and "representative," "thing" and "thought" mean then a practical distinction of the utmost importance, but a distinction which is of a FUNCTIONAL order only, and not at all ontological as understood by classical dualism.[40]

Rather than regarding consciousness or matter as a primal substance of the universe, James proposes that the one primal stuff out of which everything is composed is "pure experience"; and the function of knowing is a special sort of *relation* among components of experience. That relation itself is a component of pure experience. Thus, consciousness, the knower, the subject, or bearer of knowledge, is one "term" of pure experience; objects of knowledge are the other "term." James says that "[t]he instant field of the present is at all times what I call the 'pure' experience. It is only virtually or potentially either object or subject as yet. For the time being, it is plain, unqualified actuality, or existence, a simple *that*."[41]

While in the state of pure experience there is no self-splitting of this reality into consciousness and what the consciousness is *of*. Its subjectivity and objectivity are functional attributes solely, and they are retrospectively identified only when one re-engages with one's conceptual framework, in which subjects and objects are separated.

For James, the separation of experience into consciousness and content comes by way of addition, not subtraction. That is, a given undivided portion of experience, taken in one context of associates, plays the part of knower, of a state of mind, of "consciousness"; while in a different context the same undivided bit of experience plays the part of a thing known, of an objective "content." Thus, in one group it figures as a mental process, in another group as a mental object. And, since it can figure in both groups simultaneously, one may speak of it as being both subjective and objective.[42]

According to this theory, just as a perceptual object such as fire does not exist as an idea within an individual but is experienced outside, so is *imagined* fire not located inside of a thinking subject but occupies a definite place in the outer world. The difference, he says, between perceived and imaginary fire is that the latter cannot burn perceived sticks, though it may burn imaginary sticks. Perceived fire, which is commonly deemed "real" in comparison to imaginary fire, does have causal efficacy in the physical world; and it is on this basis that "real" experiences are distinguished from "mental" ones and things are distinguished from our thoughts of them. However, what we call the physical world consists of a confluence of these so-called real and imaginary elements. Our perceptual experiences, being the originally strong experiences, form the nucleus of this world.

A past event that is recalled in the present bears the same relation to the individual as something that is presently perceived; that is, the one is not a known object and the other a mental state. Both influence the individual in similar ways with a reality that is directly felt in experience, and both constitute one's experiential world. Even the contents of dreams and hallucinations, he points out, are still experienced "out there" and not "inside" of ourselves.[43] Just as perceived objects are present within a field of consciousness, so are recollected objects and imaginary objects states of mind; and they all demonstrate causal efficacy with respect to the mind.

While James regards both consciousness and matter as relations, and he insists that relations in general can be perceived, there is an asymmetry in his insistence that matter, but not consciousness, can be observed. The asymmetry goes further. James acknowledges that objective physical objects have their own history, composition, and effects apart from our consciousness of them; but he does not grant these qualities to mental states or to consciousness itself. Unlike physical phenomena, consciousness, he says, is composed of nothing. But this position seems peculiar if, as he proposes, consciousness is no less real than the physical phenomena that appear to it.

The asymmetry in James's view of mind and matter may be due in part to his advocacy of a "field theory" of consciousness, in contrast to an "atomistic theory," which he vigorously rejects. I would argue, however, that the nature of consciousness does not intrinsically conform either to a field theory or an atomistic theory. Rather, different kinds of conscious events become apparent when inspected from the perspective of each of these different conceptual frameworks. Using James's field theory, one may ascertain an individual, discrete continuum of awareness; and using the atomic theory one may discern within that stream of consciousness discrete moments of awareness and individual, constituent mental factors of those moments. Thus, while certain features of consciousness may be perceived only within the conceptual framework of a field theory, others may be observed only in terms of an atomistic theory. This complementarity is reminiscent of the relation between particle and field theories of mass/energy in modern physics. The crucial point here is that neither conceptual framework is inherent in the nature of pure experience. James seems to have fallen into the trap of reifying his own concept of a field of consciousness, and this may have prevented him from determining, even to his own satisfaction, the way in which consciousness does and does not exist.[44]

James did not present a practical means of transcending one's familiar conceptual framework and entering into the state of pure experience. On the contrary, he declared, "Only new-born babes, or men in semi-coma from sleep, drugs, illnesses, or blows, may be assumed to have an experience pure in the literal sense of a that which is not yet any definite what."[45] Given his keen interest in and appreciation for mystical experience, it is strange that he apparently did not consider that advanced contemplatives may have gained access to conceptually unmediated consciousness that would have a strong bearing on his notion of pure experience.

Padmasambhava on Conceptually Unstructured Awareness

According to the contemplative tradition of Padmasambhava, instead of first learning a theory of consciousness and using it to enter contemplation, one firsts seeks experiential insight into the nature of the mind, then derives one's theories from that experience. Thus, the first task is to settle one's mind in its natural state, achieve meditative stabilization, and then examine the nature of awareness.[46]

In a meditative technique taught by Padmasambhava for seeking out the nature of consciousness, one's visual gaze is steadily directed at the space in front of one. Once the awareness is stabilized, one examines that very consciousness that has become steady, and one begins questioning: Is there something real that remains clearly and steadily, or when observing consciousness, is there nothing to see? Is the one who is directing the mind and the mind that is being directed the same, or are they distinct? If they are not different, is the one that truly exists the mind that is being directed? Temporarily adopting that hypothesis, one observes: What is the nature of that so-called mind? Is it anywhere to be found among the external objects of awareness?

While steadily observing the consciousness of the one engaging in this training, one examines whether the so-called mind even exists. If so, does it have a shape? If one thinks that it may, one then examines the mind carefully to determine what that shape might be. Is it a pure geometrical form, like a sphere, a rectangle, a semicircle, or a triangle? Likewise, one examines the mind to see if it has any color or physical dimension. If one concludes that it has no such physical properties, one then proceeds to examine whether the mind might not exist at all. But if this were the case, how could something that does not exist engage in such contemplative inquiry? Moreover, if the mind is a nonentity, what is it that generates such passions as hatred? If one concludes that the mind does not exist, is there not someone or something that drew that conclusion? With this question in mind, one steadily observes whether the consciousness that ponders whether it exists is itself the mind. If it does really exist, one would imagine it must be some kind of a substance; but if so, what are its qualities? On the other hand, if it does not exist, who or what is it that thinks this? In this way one's awareness is drawn inward, grappling with and breaking down the conceptual constructs of existence and nonexistence with respect to the mind.

In such introspective inquiry one also examines the origins, location, and disappearance of mental phenomena. One examines, for example, whether mental events arise from the external environment or from the body; and one investigates the exact manner in which they arise moment by moment. Once they occur, one investigates where they are present—whether outside or inside the body—and if they seem to be present inside the body, one examines exactly where they are located.

In addition, one inquires whether the mind and thoughts are the same or different. At times the mind is withdrawn from appearances and seems to be empty, and at times it engages with phenomena. Are those appearances and that emptiness the same or distinct, and are the stillness and the activities of the mind the same or different? If they are distinct, when does this differentiation occur, and what is the demarcation between them? Finally, when thoughts and other mental events cease, how does this occur? Do they proceed from existence to nonexistence, or do they go somewhere beyond the field of consciousness? If they do depart, do they leave in the same aspect as the one in which they were previously present, or do they depart in a more ethereal manner?

Padmasambhava comments on the results of such inquiry as follows:

given the differences in intellect, in some, a nonconceptual, unmediated, conceptually unstructured reality will arise in their mind-streams. In some there will be a steadiness in awareness. In some, there will be a steady, natural luster of emptiness that is not an emptiness that is nothing, and there will arise a realization that this is awareness itself, it is the nature of the mind. In some, there will arise a sense of straightforward emptiness. In some, appearances and the mind will merge; appearances will not be left outside, and awareness will not be left inside. There will arise a sense that they have become inseparably equalized.[47]

At this point, one's mentor is to offer the following guidance:

once you have calmed the compulsive thoughts in your mind right where they are, and the mind is unmodified, isn't there a motionless stability? Oh, this is called "quiescence," but it is not the nature of the mind. Now, steadily observe the very nature of your own mind that is being still. Is there a resplendent emptiness that is nothing, that is ungrounded in the nature of any substance, shape, or color? That is called the "empty essence." Isn't there a luster of that emptiness that is unceasing, clear, immaculate, soothing, and luminous, as it were? That is called its "luminous nature." Its essential nature is the indivisibility of sheer emptiness, not established as anything, and its unceasing, vivid luster. Such awareness is resplendent and brilliant, so to speak.[48]

Such conceptually unstructured awareness, Padmasambhava claims, does not originate at any specific time, nor does it arise from certain causes and conditions. Likewise, such awareness does not die or cease at any specific time. While it does not conform to our notion of existence, its unimpeded creative power appears in all manner of ways, so it is no one single thing. On the other hand, while the mind takes on many different appearances, it has no inherent nature of its own, so it is not a multitude of things either.

Thus, Padmasambhava thoroughly rejects the hypothesis that consciousness is some kind of purely subjective, spiritual substance from which all mental phenomena emerge.[49] In this way, one is said to come to a conceptually unmediated experience of the nature of unmodified awareness.

Awareness, Padmasambhava suggests, is like a wild stallion that has roamed freely for so long that its owner cannot recognize it. It is not enough for the horse to be pointed out to its owner; rather, once that has been done, methods must be used to capture it, train it, and put it to work. Likewise, it is not enough merely to identify the nature of one's own wild mind; one must now sustain and utilize the experience of conceptually unstructured awareness. Now, as before, one's visual gaze and mental awareness are fixed in the space in front of one; and without meditating on anything, one lets the mind come to rest steadily, clearly, homogeneously, and without wavering.

At the beginning, one practices for only short sessions, but as one becomes accustomed to the training, the duration of each session is increased. When bringing each session to an end, one slowly emerges from contemplation without losing the sense of unstructured awareness, without distraction, and without conceptual grasping. Such unwavering mindfulness is to be maintained during all activities of eating, drinking, speaking, moving, working, and so on. Apart from that, there is nothing on which to meditate, for the introduction of any artificial technique into such experience only obscures the conceptually unstructured nature of pure experience.

Following each session of contemplation, whatever ideation arises, one repeatedly lets it appear and vanish of its own accord, without grasping onto it or its intentional object. When a hateful thought arises and later gives way to a compassionate thought, the earlier hatred did not go anywhere but is released by itself. Hatred never remains immutably. Moreover, according to this Great Perfection tradition, all mental processes, even afflictive ones such as hatred, are natural displays of the creative power of pristine, conceptually unstructured awareness. From this perspective, hatred and other mental processes are seen to be unborn, having no location or real existence of their own.

Conceptually unstructured awareness—which is nondual from the phenomena that arise to it—is regarded as the ultimate reality, and the realization of such nondual consciousness is the final goal of contemplative practice.[50] In this experience, the very distinction between public, external space, in which physical phenomena appear to occur, and private, internal space, in which mental phenomena appear to occur, dissolves into a "mysterious space," which is the very nonduality between the conceptually constructed external and internal spaces. The ultimate nature of objective phenomena, therefore, is found to be none other than the ultimate nature of subjective phenomena; and that is the nonduality of appearances and awareness. When one achieves perfect realization of this state, in which there is no longer any difference between one's awareness during and after formal meditation sessions, it is claimed that one's consciousness becomes boundless in terms of the scope of its knowledge, compassion, and power. Hence, the contemplative pursuit of such realization is said to be the most sublime of sciences.

The Spectrum of Consciousness

While most contemporary Western scholars of religion claim that all mystical experiences are conceptually structured by one's beliefs, memories, expectations, and desires,[51] the Buddhist and many other contemplative traditions throughout the world claim that conceptually unmediated, "pure" consciousness is indeed a possibility. Many report that in such experiences one's sense of one's own independent, subjective, consciousness vanishes together with one's sense of independent, objective objects of consciousness. Thus, the very experience of the duality of subject and object collapses.[52]

The distinguished scholar of mysticism Robert K. C. Forman characterizes such a state of pure consciousness as "a mind which is simultaneously wakeful and devoid of content for consciousness,"[53] and he claims that virtually identical techniques for achieving such a state have been practiced in many disparate contemplative traditions throughout history. Common elements to all these practices is withdrawing the attention inward, away from all sensory and conceptual stimuli, together with a conscious "forgetting" of one's language and concepts.

If we were to be satisfied with this twofold classification of "pure" versus "impure" consciousness, however, many of the subtle gradations of conscious experiences would be overlooked. For example, the mere fact that one has temporarily disengaged one's attention from all words, thoughts, and mental contents does not necessarily imply that one's experience is no longer at least subliminally structured by one's conceptual framework. The well-known processes of "precognitive structuring" and "subliminal priming" of experience are bound to play a role in most, if not all, conscious states. Thus, the experiences of two people with different backgrounds who enter such a state of seemingly pure consciousness may be significantly different, as may be the residual, lingering effects of their experiences.

Rather than postulating a simple dichotomy of pure and impure states of consciousness, it may be more useful to consider a spectrum of conscious states, ranging from those that are highly structured by one's language and concepts to those that are less structured. This spectrum is akin to Hilary Putnam's theory of a continuous spectrum of subjective/objective statements and perceptions of reality. Toward one end of this spectrum would be experiences that are purely fabricated by one's expectations and beliefs; at the other end would be experience that is utterly free of such subjective influences. Indeed, utterly pure consciousness may be uninfluenced even by one's physical senses, which are specifically human in nature.

Scientists and contemplatives alike are challenged to distinguish between their conceptual superimpositions upon experience and the actual evidence that is being presented to their senses, including the sense of mental perception. But as long as scientists are focusing their attention outward, there may be little possibility of their entering a state approaching pure consciousness; and this is even less likely as long as they are viewing the objective world through the subjective filters of their scientific concepts. Many contemplatives, on the other hand, seek to disengage from their conceptually structured experiences derived from both sensory and mental perception and to enter a state free of all subjective constructs. The earlier discussion of Buddhist techniques for withdrawing the attention into the nature of consciousness itself illustrates prime examples of this pursuit. However, many Buddhist contemplatives have been quite aware of the common error of mistaking such a conscious state for one that is utterly unstructured by language and concepts. With this recognition, contemplatives such as Padmasambhava have devised further contemplative methods for "breaking through" *all* conceptual mediation to a state of primordial awareness that transcends specifically human consciousness itself.[54] But in transcending the pole of human subjectivity, they simultaneously transcend the pole of human objectivity in a state that is simply nondual.

Within the spectrum of consciousness structured to varying degrees by one's beliefs, desires, and expectations, delusional modes of experience tend to be toward the highly structured end of the spectrum, and they are commonly accompanied by intense misery. They are said to be out of touch with reality, and individuals suffering from such delusions are commonly given psychological therapy to enable them to distinguish their conceptual fabrications from reality. In the midrange of the spectrum are located our common everyday experiences of the world of subjects and objects, as well as the entire range of empirical and theoretical scientific research; and these are commonly accompanied by transient joys and sorrows. Within this spectrum of experience, if one attends primarily to objective phenomena, those phenomena seem more real; whereas if one attends primarily to subjective phenomena, those phenomena seem more real. This is simply a matter of one's beliefs and interests.

Many contemplatives claim that at the far end of the spectrum of consciousness—furthest removed from states of pathological delusion—there is a state of awareness that utterly transcends all conceptual constructs, including the dualities of subject/object, existence/nonexistence, self/other, and mind/matter; and this state is widely reported by contemplatives to be imbued with an unprecedented, enduring, great bliss. Countless contemplatives in diverse cultures claim to have realized such a state, and they declare that the state itself, together with its lasting, residual effects, brings with it the highest knowledge and the greatest value. For in this state of genuinely pure consciousness, which is inconceivable and ineffable, one realizes that which is truly ultimate. Whether or not such extraordinary claims are valid is a question that may forever elude conceptual analysis or argumentation, but it may possibly be answered through one's own experience.

The range of contemplative experiences may be highly significant not only for religious people but for the scientific exploration of consciousness. As physicist and Nobel laureate Brian Josephson comments,

> [i]n the case of conscious experience . . . simply specifiable states of consciousness exist. Typically, these states consist of what may be called "pure" ideas or emotions. Most basic of all is the state known as pure consciousness or samadhi, which has no identifiable content other than being conscious . . . Pure consciousness is that limiting state of consciousness which is completely undisturbed by other entities; in other words it consists only of the phenomenon of consciousness interacting with itself.[55]

Physicist Evan Squires responds to this assertion as follows.

> In principle, it appears likely that the study of consciousness through these types of activity could help us in understanding its true nature. . . . If, or when, we ever have a science of conscious mind, there is little doubt that states of contemplation and of dreaming, etc., will play a big part in the experiments we do. Maybe then we will understand them better than we do at present.[56]

Is it possible that contemplative experiences that transcend our ordinary reified concepts of subject and object and so on may have a strong bearing on the insights drawn from modern physics itself? Physicist Nick Herbert comments that the source of all quantum paradoxes appears to lie in the fact that human perceptions create a world of unique actualities—our experience is inevitably "classical"—while quantum reality is simply not that way at all. And he asks, "Since physics assures us that our lives are embedded in a thoroughly quantum world, is it so obvious that our experience must remain forever classical?"[57] Perhaps our experience can transcend the confines of our everyday, dualistic, "classical" concepts. If so, a contemplative science of consciousness may lead the way out of our present confusion concerning the nature of consciousness and its relation to the rest of the natural world.

PART III

The Resistance

Dismiss whatever insults your soul.

Walt Whitman,
Leaves of Grass

THE MIND IN SCIENTIFIC
MATERIALISM

The Scientific Banishment of Subjectivity

In the dualistic, mechanical philosophy that dominated the rise of modern science, nature was not only seen as devoid of consciousness but also was objectified to the point that it was divorced from perceptual experience altogether. The material objects that made up the world were believed to have certain primary qualities, such as size, shape and velocity; but they were inherently devoid of all secondary properties, such as color, smell, and sound, which were relative to perception. Thus, conscious experience was effectively removed from nature and, therefore, from the objective domain of science.

As the scientific worldview developed, words that previously referred to constituents of human sensory experience were defined in purely objective terms. Sound became fluctuations in an objective medium such as air; smell became molecules adrift in the atmosphere; light became a form of electromagnetic energy; and color became specific frequencies of that energy. Science was concerned solely with these phenomena as they were thought to occur independently in nature. Adhering to the principles of scientific materialism, science came to be equipped with more and more sophisticated means of exploring objective physical processes; but there was no corresponding development of means to explore subjective cognitive processes. Thus, scientists simply redefined secondary properties—such as color, sound, and so on—in terms of the objective physical stimuli for the corresponding subjective experiences. In so doing, they shed increasing light on the nature of these physical phenomena, while shedding little or no light on the corresponding subjective perceptions. Thus, subjective experience was not explained; rather, it was overlooked through a purgative process of objective redefinition.

Only in the late nineteenth century did a science of conscious states of experience begin to emerge. But, as noted earlier, by the early twentieth century, American academic psychologists had shifted their attention away from subjective states of consciousness to the objective study of behavior. Behaviorists were not so much concerned with redefining terms of subjective experience in purely objective terms as they were in ignoring subjective phenomena and terminology altogether.

The more traditional mode of purging subjectivity from the natural world returned in the late 1950s, with the emergence of cognitive psychology. This new discipline grew out of the psychology of information-processing as well as communication and control theory. *Information* now became added to the long list of cognitive terms that were to be purged of their subjective content. According to this new sanitized interpretation, information is simply a decision between two equally plausible alternatives, independent of any specific content. In the process, the notion of a decision also came to be viewed as a purely objective, mechanical process. In this scientific context, the transmission and reception of information, like the process of scientific observation and measurement discussed earlier, is thought to occur without reference to any conscious, subjective agent. While in common speech *information* is intimately related to the notion of *meaning*, this in no longer the case in scientific usage.

If information is objectively embedded in physical phenomena—perhaps in *all* such phenomena—without reference to any subjective agent, this would imply that gibberish and articulate speech or a book filled with random scribbling and a biology textbook, are equally informative. Moreover, individuals using different languages and conceptual frameworks should also have equal access to the information that exists intrinsically in objective phenomena. But I would argue that the information within a computer exists only because it has been *put there* by a conscious, subjective agent. It exists by consensus among people for whom certain symbols have been designated to convey meaning. Apart from that put by conscious computer-users, no information is loaded into, stored, processed, or produced by computers. Similarly, while the brain is described as an organic computer in which symbols and information are received, stored, and processed, none of this could happen without relation to a conscious brain-user or brain-observer. So we are left with the questions: if the brain processes information, who put it there, who is responsible for the input, and who experiences the output?

If we combine scientific materialism's objective interpretation of observation and measurement with the objective interpretation of decision-making and of conveying and receiving information, it would follow that the entire universe is constantly involved in observing itself, communicating with itself, and making decisions about the information it sends and receives. Cognitive terminology is now used in reference not only to computers and brains but to photons and a wide range of other nonorganic phenomena. Thus, scientific materialism has been ushering in a new version of anthropomorphic animism, in apparent opposition to the traditional scientific enterprise of depersonalizing the physical world. We now have cases of scientists working in the field of quantum mechanics "observing" photons "making choices"; neuroscientists "seeing" thoughts, perceptions, and emotions in computer-generated images of the brain; and computer scientists "witnessing" their creations, like modern-day Pinnochios, thinking, remembering, solving problems, and learning. It seems that believing is seeing, much as many people in sixteenth-century Europe observed the paranormal feats of witches, and animistic societies throughout history have seen mountains, lakes, and trees to be inhabited with a myriad of spiritual entities.

It is hardly any wonder, then, that panpsychism—the belief that all matter throughout the universe is conscious—is being advocated by an increasing number of scientific materialists. But pragmatically speaking, there seems to be little difference between asserting that *everything* is conscious and that *nothing* is conscious. Neither view provides a compelling explanation of the nature, origins, or function of consciousness in beings we *know* are conscious, such as ourselves.

There is one major difference between the animism of scientific materialism and that of many earlier nature religions: traditional nature religions personalized the universe by believing that external physical objects, much like humans, are imbued with spiritual forces separate from matter. In this way, the gap between personal and impersonal phenomena was narrowed. Scientific materialism narrows this gap in the opposite way by asserting that the cognitive processes commonly associated with humans actually occur throughout nature; and in all cases they are impersonal processes devoid of any spiritual forces separate from matter. Thus, while appearing to personalize nature, scientific materialism actually depersonalizes human existence.

From a scientific perspective, the advantage of this redefinition of information is that one can now focus on the efficacy of any communication of messages via any mechanism; and this made it possible to consider cognitive processes apart from any particular embodiment, such as a human subject. Cognitive psychology was founded on the assumption that, for scientific purposes, human cognitive activity must be described in terms of symbols, schemas, images, ideas, and other forms of mental representation, without any necessary connection to consciousness. However, psychologist Howard Gardner comments, this procedure resulted in "a huge gap between the use of such concepts in ordinary language and their elevation to the level of acceptable scientific constructs."[1] Just as such perceptual terms as *color* and *sound* had earlier been objectified, modern cognitive science has now redefined most subjective cognitive terms so that they, too, are purged from the subjectivity of everyday conscious experience.

The modern field of cognitive science has come to include a diverse range of disciplines, including the neurosciences, artificial intelligence, philosophy of mind, psychology, linguistics, quantum theory, and evolutionary theory. In many of these disciplines the computer has become the central mechanical model of the mind and cognition is identified with symbolic computations. Thus, cognitive science becomes the study of such cognitive symbolic systems, and the field of artificial intelligence takes this cognitivist hypothesis literally. Gardner acknowledges that one of the principal features of cognitive science is the deliberate decision to de-emphasize certain factors that may be important for cognitive functioning but whose inclusion would unnecessarily complicate the cognitive scientific enterprise. As it turns out, the de-emphasized features are those that do not conform to the computer model, which, he claims is "central to any understanding of the human mind."[2]

During the Scientific Revolution, some natural philosophers likened the mind to a hydraulic system, and an early twentieth-century metaphor for the mind was a telephone switchboard. Regardless of how fundamentally dissimilar the mind is to the latest products of technology, including the modern computer, scientific materialists have long been convinced that it must be similar to some kind of ingenious, material gadget. The most salient omission in this regard is consciousness itself, but it is now commonly presumed that consciousness really boils down to nothing more than information-processing.[3]

Scientific materialists have long sought to sublate the existence of subjective phenomena by reducing them to objective phenomena. For example, by identifying the causal role of photon emissions of 600 nanometers in producing the experience of red, scientists have commonly reduced red to such photon emissions. In other words, "red" is simply redefined so that it can be identified with the light reflectances that are its objective, physical cause, which are now presented as what "red" really is. By so doing, red as we perceive it is not eliminated; it is simply no longer called "red." And the experience of red is not explained through such a reduction; it is simply ignored. Similarly, if conscious states such as pain are ontologically reduced to patterns of neuron firings, the essential features of those phenomena are not explained; they are simply left out.

Because the phenomena of sensory and mental experience have been excluded from the world of scientific materialism, some philosophers have tried to retrieve these contents of experience by designating them "sensory data" or "qualia." But each of these words has been variously defined in such complex, highly abstract, philosophical terms that the very existence of such phenomena is frequently called into question, particularly by scientific materialists. Thus, modern philosophers who do not embrace scientific materialism find themselves in the bizarre situation of needing to coin a new term to designate the experienced objects of consciousness. In the following discussion I shall use the term *qualia* simply to denote both the immediate contents of sensory and mental awareness—such as physical pain, emotions, mental imagery, and perceived colors, sounds, smells, and so on—as well as the subjective consciousness of these phenomena. I hope this definition makes it obvious that qualia in this sense of the term do exist and that to ignore them is to ignore firsthand experience altogether.

Why have physical scientists sought to reduce such qualia to their physical causes? The most plausible answer is that, due to the dictates of scientific materialism, their interests have been directed solely to those objective phenomena. If neuroscience had developed before the science of optics, color might well have been reduced to processes in the visual cortex instead of frequencies of electromagnetic radiation. Thus, the explicitly causal and implicitly ontological reduction of conscious experience to its physical causes is a traditional way of excluding subjectivity from the natural world *because of lack of interest*. But if we *do* take an interest in the nature of conscious states themselves, it seems we must look elsewhere than the present methodologies of the sciences, which have not been able to account for them successfully.

Ironically, one of the discoveries of recent neuroscientific studies of perception is that there is no one-to-one relationship between light flux at various wavelengths and the colors we perceive things to have.[4] Thus, since perceived colors have no direct physical counterparts, we cannot account for our experience of color as an attribute of things in the world by appealing simply to the intensity and wavelength composition of the light reflected from an area. This has direct implications for visually perceived objects in general, since it is contrast and borders that visually form those objects.

It remains an open question whether specific sensory or mental phenomena will in fact turn out to have one-to-one relationships with external physical events or with specific internal brain functions. This raises an interesting array of questions concerning all such qualia: If they do not exist in the outside world or inside the head, where, if anywhere, do they exist? How do the symbolic expressions purportedly encoded in the brain get their meaning? How are we to distinguish between the conscious and unconscious transmission and reception of those symbols? Thus far, the neurosciences have largely overlooked these issues, and until recently the terms "consciousness" and "awareness" did not even appear in most of the textbooks and dictionaries for any of the cognitive sciences.

The Identity of Mind and Brain States

Regardless of any scientific evidence concerning the mind/brain relationship, the dictates of scientific materialism predetermine that there are only two legitimate ways of coming to terms with subjectively experienced mental phenomena: (1) either they are actually physical phenomena or properties of matter, or (2) they do not exist at all. In the early days of modern neuroscience, scientific materialists declared that there are no such things as separate mental phenomena, because in reality *they are identical with brain states.* As the one-to-one correspondence between specific mental processes and specific brain processes appeared more and more dubious in light of neuroscientific advances, scientific materialists declared that there are no such things as separate mental phenomena, because in reality *they are not identical with brain states.* Evidently, this ideology finds it necessary to get rid of subjective mental phenomena, regardless of the scientific evidence.

Even the most ardent scientific materialists acknowledge that we do not presently know enough about the intricate functioning of the brain to establish the equivalence of specific, subjective mental processes with specific, objective brain processes. Nevertheless, physicalists such as Patricia Churchland maintain that we know *too little* about the brain at present to conclude that consciousness is a product of anything other than brain functions. So, instead of drawing on scientific *knowledge* to argue that consciousness is solely a product of the brain, such people argue that we should accept their belief on the basis of scientific *ignorance*! This line of reasoning is similar to that used by Christian theologians who have attributed to divine influence those natural phenomena that are not yet explained by science. This notion of the divine has been dubbed the God-of-the-gaps, suggesting a new term, "materialism-of-the gaps," to characterize this strategy used in support of scientific materialism.

From the perspective of this belief system, the burden of nonequivalence between the two rests with dualists, for there is no doubt at all that physical matter exists, while the notion of mental phenomena is seen as being nothing more than a tenuous hypothesis. Thus, it is up to opponents of scientific materialism to demonstrate that such a physical reduction is outright impossible.[5] For those who are not advocates of this ideology, however, this may seem like a peculiar demand. For each of us as human subjects, what is more real than our joys and sorrows, hopes and fears, desires and beliefs, and our sensory experience of the world about us? On what grounds are we to believe that these mental phenomena are any less real than such physical phenomena as mountains and buildings, let alone quarks and electromagnetic fields?

Clinical neurology has indeed gained remarkable insights into functional parts of the human brain without which mental states do not arise or are altered in some demonstrable ways. However, while these findings tell us that in the absence of those specific regions of the brain, specific mental states are systematically, categorically altered, these discoveries do not tell us what those regions do by way of causing mental states. Neuroscientists, too, have begun to identify some of the brain correlates to specific states of human consciousness. For instance, Francis Crick and Christof Koch believe that consciousness depends crucially on some form of serial attentional mechanism that helps sets of the relevant neurons to fire in a coherent semioscillatory way, probably at a frequency in the 40–70 Hz range. In this way, a temporary global unity is imposed on the neurons in many different parts of the brain. Their findings indicate that consciousness is correlated with a special type of activity or perhaps a subset of neurons in the cortical system, and they hypothesize that there is one (or perhaps a few) basic mechanism that underlies all the different forms of consciousness.[6]

But what, precisely, is the nature of the correlations between such brain processes and mental processes? Research in this area, most notably by neuroscientist Benjamin Libet, reveals that there is commonly a time lag of approximately one-tenth of a second between a brain process and its corresponding mental process. This would suggest a *causal* relationship, rather than an *identity* relationship between the two. This issue, however, is complicated by the fact that other brain processes are occurring simultaneously with the mental process under examination; and it is difficult to determine whether a specific mental process is more closely associated with that particular brain state or one that precedes it.

Assuming that mental processes are a function of brain processes, and assuming that correlates are discovered between specific brain states and specific mental states, the exact nature of that correlation remains open to interpretation. William James proposed three feasible theories to account for such correlations: (1) the brain produces thoughts, as an electric circuit produces light; (2) the brain releases, or permits, mental events, as the trigger of a crossbow releases an arrow by removing the obstacle that holds the string; and (3) the brain transmits thoughts, as light hits a prism, thereby transmitting a surprising spectrum of colors.[7] Among these various theories, the latter two allow for the continuity of consciousness beyond death. James, who believed in the third theory, hypothesized that

> when finally a brain stops acting altogether, or decays, that special stream of consciousness which it subserved will vanish entirely from this natural world. But the sphere of being that supplied the consciousness would still be intact; and in that more real world with which, even whilst here, it was continuous, the consciousness might, in ways unknown to us, continue still.[8]

If the brain simply permits or transmits mental events, making it more a conduit than a producer, James speculated that the stream of consciousness (1) may be a different type of phenomenon from the brain, (2) interacts with the brain while we are alive, (3) absorbs and retains the identity, personality, and memories constitutive in this interaction, and (4) can continue to go on without the brain. Remarkably, empirical neuroscientific research thus far is compatible *with all three hypotheses* proposed by James, but the neuroscientific community on the whole has *chosen* to consider only the first hypothesis, which is the only one compatible with the principles of scientific materialism. Thus, instead of letting empirical evidence guide scientific theorizing, a metaphysical dogma is predetermining what kinds of theories can even be considered, and therefore, what kinds of empirical research are to be promoted.

This kind of dogmatically driven adjudication of reality is more commonly associated with religion than with science. For example, some sixteen hundred years before James, Augustine also pondered the origins of the human mind and soul within the parameters of Christian theology. After careful biblical research, he presented the following four hypotheses: (1) an individual's soul derives from those of one's parents; (2) individual souls are newly created from individual conditions at the time of conception; (3) souls exist elsewhere and are sent by God to inhabit human bodies; and (4) souls descend to the level of human existence by their own choice.[9] After asserting that all these hypotheses may be consonant with the Christian faith, he declared: "[i]t is fitting that no one of the four be affirmed without good reason."[10] The problem of the origin of the soul remained unsolved to the end of Augustine's life, as it does for Christian theology today. While Christians commonly attribute the origin of the soul to their ultimate reality, God, scientific materialists, who are equally ignorant of the origin of consciousness, attribute it to their ultimate reality, matter.

The nature of mind/brain correlations is especially difficult to determine with a high degree of precision because of the difficulty of detecting exactly when a mental process occurs. Some neuroscientists believe they are moving toward the invention of a *psychometer* that could do just that, while others more cautiously assert that such a device could at best detect the physical correlates that invariably accompany specific conscious states and possibly the presence of consciousness itself. Such a psychometer, which is being devised using sophisticated brain-imaging techniques, would detect the physical indicators of what one is experiencing as one is experiencing it. Champions of this approach believe that such devices could indirectly detect mental events in a manner similar to other systems of measurement designed to detect other hidden processes such as quantum events. Could such a psychometer detect the presence or absence of consciousness in reptiles, insects, or microorganisms or produce empirical evidence to settle questions concerning plant or mineral consciousness? An optimistic response is that once we have reliable correlates of specific mental contents such as perceiving and memory for humans—where first-person access to those subjective states is possible—there is no reason not to apply those correlations from one species to another, at least in higher animals.

While some neuroscientists are seeking the neural correlates of conscious and unconscious mental states in humans, others are trying to identify the structural conditions over the course of evolution under which the first intentional capacities in living organisms could emerge. Research in this field is focused on identifying objective measures of the most basic forms of cognition in primitive organisms. The rationale for this approach is that primitive animals developed cognitive capacities, including memories, imagination, and communication skills, long before humans did. One can then ask: Are the most basic forms of cognition truly *conscious*, in the sense of the animal actually perceiving its environment or its own physical presence? Or is such basic cognition nonconscious, but nevertheless intentional, in the sense that the organism has some kind of nonexperiential knowledge of incoming stimuli and motor skills? To answer this question, one would have to have a true psychometer that could distinguish between conscious and nonconscious cognition. Perhaps there will be such a device that can detect these differences in humans, but it is not clear how much light such research can shed on consciousness and cognition in other mammals, let alone in insects (which have no cortex), plants (which have no central nervous system), or minerals.

While it is tempting to assume that certain behavior in primitive organisms is intentional because of its similarity with human behavior, such conclusions are tenuous. Shall we conclude, for example, that insect-eating plants display intentional behavior on the grounds that some humans eat insects? Shall we conclude that sophisticated computers perform intentional behavior since humans, too, remember and process information? Can we say that sophisticated robots are engaging in intentional behavior when they perform complicated tasks that humans also perform? Or can we assume with confidence that people who are asleep or even in a coma are utterly devoid of consciousness or cognition since they are inactive?

It is equally problematic to infer the presence of consciousness, intentionality, or even the most primitive forms of cognition purely on the basis of objective behavior. This is not to argue that other mammals, insects, or even microorganisms are devoid of cognition or consciousness; the point is rather that the use of any psychometer now under development is unlikely to provide sufficient empirical evidence for answering such questions.

Evidently, psychometers as they are presently envisioned by neuroscientists will not be able to detect whether a specific brain correlate *precedes* and *causally produces* its corresponding conscious state or whether the conscious state is actually an *emergent property* or *function* of the neural correlate, which would therefore occur *simultaneously* with it. To make that temporal distinction, one would need an instrument that could actually observe the precise moment of the origination of the conscious state. That is, one would need a real psychometer, as opposed to a neural correlate detector.

In this regard, there is another distinction between brain states and mental states that merits attention. Neuroscientists may detect the precise brain correlates of a perception of red, for example, and they may be able to manipulate brain processes directly to induce such a perception. But the red one sees when observing a rose and when the brain is so manipulated are not the same: the first perception is a valid cognition of the color of something external to the brain; and the second is a hallucination created by an internal manipulation of the brain, with no objective referent. It makes sense to speak of valid and invalid perceptions but not of valid and invalid brain processes. Likewise, neuroscientists may manipulate the brain so that one feels pleasure. But they cannot manipulate the brain to arouse *meaningful* joy, such as the joy in successfully warning people of an approaching tornado, so that no lives are lost. For such joy is a response to experienced events *external* to the brain. If it were possible to arouse a *sense* of meaningful joy purely through brain stimulation, one would have to conclude that this sensation was delusive. Similarly, we now have a wide array of drugs that induce *sensations* of mystical experiences, but whether these are actually identical to genuine experiences of the divine is certainly open to question. While there are valid and invalid cognitions, there are no valid and invalid brain states, any more than there are meaningful and meaningless brain states.

While neuroscientists may identify specific neurophysiological events that arise as correlates to specific conscious states, it is not apparent what that would tell us about the nature or the origins of consciousness. Some researchers immediately proceed to *define* consciousness in terms of those correlates, thereby reducing the whole range of conscious states to those correlates, and then conclude that the problem of consciousness has been solved. Even if neuroscientists are eventually able to map every single neural mechanism and identify its mental correlate (where such exists), all that will be established is that there are neural correlates for mental phenomena; but such knowledge won't resolve the debates concerning physicalism, dualism, and the general mind/body problem.

The present state of the Western neuroscientific and philosophical study of consciousness seems eerily similar to the medieval scholastic study of astronomy. Neuroscientists and neurophilosophers are steadfastly studying the brain, which they are sure is solely responsible for the production of all conscious states, much as medieval monks studied the Bible, with the certainty that its Author was solely responsible for the creation of the universe. And just as medieval theologians spun out elaborate theories of the heavens based on the Bible, Aristotle, and Ptolemy, so are present philosophers devising a myriad of mutually incompatible theories of consciousness, without establishing much consensus among themselves.

What medieval astronomers lacked was methods for making more precise and thorough observations of heavenly bodies themselves, which would then form the basis for devising an empirical science of astronomy. The conceptual blockage that was holding them back was a set of metaphysical beliefs that made them think such empirical methods were not necessary. What modern cognitive scientists lack is methods for making more precise and thorough observations of subjective states of consciousness, which would then form the basis for devising an empirical science of consciousness. The conceptual blockage that is holding them back is a set of metaphysical beliefs that make them think such firsthand, empirical methods are not necessary. While neuroscientists are making wonderful strides in exploring the brain and its relation to the mind, when it comes to exploring the nature of consciousness itself, their research is inhibited by ideological taboos.

The modern study of consciousness is long overdue for a radical re-evaluation of its most cherished assumptions and for the development of radical new methodologies for empirically investigating the nature of consciousness. As philosopher Güven Güzeldere comments in the conclusion of his overview of Western studies of consciousness, "[i]f anything, the survey of the contemporary issues and current debates surrounding consciousness points to a need for a careful re-examination of our pre-theoretical intuitions and conceptual foundations on which to build better accounts of consciousness."[11]

Mind As a Property of the Brain

According to the official scientific interpretation of evolution taught in all major textbooks of American colleges, there are utterly compelling grounds for concluding that consciousness is an emergent property of matter. All one needs to come to this conclusion is to accept a few basic facts concerning the evolution of life on earth: (1) long ago, after the formation of the earth from hot geological times, there were only nonliving molecules; (2) the first forms of bacterial life on earth emerged from these molecules; and (3) evolution proceeded from bacteria to humanity. The inevitable conclusion that must be drawn from these facts, this argument runs, is that consciousness can be nothing but an emergent property of matter.

To place this argument in context, we should recall that science has ways of empirically detecting physical phenomena such as molecules, but it has never had ways of scientifically detecting subjective phenomena such as consciousness. To emphasize a crucial point: if all we had to rely on for our knowledge of the universe were the theoretical and empirical tools of science, we wouldn't even know that consciousness exists in the universe. It is difficult to comprehend the profound limitations of this lopsided pursuit of knowledge of the natural world. To know so much about the objective world and to know so little (scientifically) about the subjective world (and nothing at all about consciousness) sets the stage for a terribly biased view of existence in which subjectivity in general and consciousness in particular are doomed from the outset to have at most a peripheral role in nature. Given the metaphysical axioms of scientific materialism, which have guided scientific research in evolution, how could subjective events ever have been regarded as anything *but* epiphenomena of material processes that scientists *do* know a lot about?

While the first *forms* of bacterial life may well have emerged from non-living molecules (where else would their bodies have emerged from?), and while there is compelling evidence to suggest that evolution proceeded from bacteria to man, scientists do not know enough to count out the possibility of nonphysical influences in the emergence of life and consciousness on earth. They are simply ill prepared to consider such a possibility, let alone explore it empirically, because of the ideological and methodological constraints of scientific materialism.

A necessary implication of the metaphysical assumptions of scientific materialism is that all mental phenomena are emergent properties or functions of matter. The practical effect of this is the prioritization of objective physical phenomena over subjectively experienced mental phenomena, which implies that research into the former will be better funded and hold higher status than research into the latter. Whenever we designate one phenomenon as being an emergent property of another, this implies an asymmetry between the two: the emergent phenomenon is secondary, and that which gives rise to it is primary. This implies, in turn, that the influence of the primary upon the secondary is greater than the influence of the secondary upon the primary; indeed, the secondary may have no causal efficacy of its own; whereas the primary does have causal efficacy of its own independent of the secondary. In short, the primary is more real than the secondary, which implies that more human resources should be directed to understanding and learning how to control the primary than the secondary.

What is the nature of matter, which, according to scientific materialism, gives rise to consciousness solely by way of the complexity of its organization? In their assertion of the primacy of matter, scientific materialists are primarily relying upon their sensory *perceptions* of matter in the macroworld, which do not exist independently in the objective, physical world and are not observable in the brain. Nevertheless, the perceptions of those attributes are regarded as emergent properties of complex configurations of atoms that exist independently of our sensory perceptions.

What then is the nature of atoms and their fundamental elementary particles? Despite the consensus among physicists concerning the mathematical laws that account for observable phenomena arising due to atomic interactions, they differ considerably in their accounts of what atoms *are*. Physicist Bernard d'Espagnat, for example, maintains that atoms are emergent properties of space or space-time, but astronomer Edward Harrison points out that there are countless possible spaces with their own geometries, and all are equally valid and self-consistent. Werner Heisenberg declares that atoms are not *things* at all, and Henry Stapp claims that elementary particles are not independently existing, unanalysable entities, but rather sets of relationships.[12] As for the actual nature of energy existing in the objective world of nature, recall Richard Feynman's comment that physicists today "have no knowledge of what energy *is*."[13] When we move our attention away from perceptual qualia associated with matter and energy to what exists independently of our perceptions, we move into a realm of *conceptual* qualia. These ideas, including mathematical formulas, do not exist independently in the objective physical world and are not observable in the brain. It seems, therefore, that all of our understanding of matter consists of dependently related events in which the subjective element is never totally absent.

To extend this line of reasoning, organic, biological events are regarded as emergent properties or functions of complex configurations of inorganic chemical processes, and those, in turn, as emergent properties or functions of complex configurations of atoms and elementary particles. Thus, the many manifestations of matter in the macro-world of everyday experience—including the experiences of neuroscientists observing the brain—are all emergent properties or functions of complex configurations of elementary particles. But when it comes to the most fundamental elements of the physical world, both mass and energy appear so abstract, insubstantial, and nonlocal that they seem to be more like conceptual constructs than minute, discrete chunks of real matter existing in some objective world independent of any system of measurement, observation, or conceptual framework. According to this analysis, all our perceptions and conceptions of matter are emergent properties or functions of the mind, which brings this process of emergence to full circle. In this case, neither mind nor matter is primary or independently real and substantial. Rather, all mental and material phenomena arise as dependently related events, with none of them being more real or primary than any other. If this is true, it would be appropriate to broaden the range of scientific methods for exploring not only objective physical events but subjective mental events, so that the interdependence between the two may be better understood.

As noted previously, scientific materialism assumes that all mental states and behavior are completely determined by the brain and its physical interaction with its environment. The logical inference drawn from this is that mental processes are in fact nothing more than physical properties of brain processes.[14] A subtle variant of this theme asserts that the actual nature of mental processes is to be found in the *functional relations* of a physical entity such as a brain, a computer, or some other instrument of artificial intelligence. According to this view, mental states are not defined in terms of either their own intrinsic, experienced properties or the intrinsic properties of brain states. Rather, they are defined in terms of the *functions* of brains and other intelligent physical systems.

The chief shortcoming of this interpretation is that scientists have no explanation as to *how* or *why* specific functions of the brain correspond to states of consciousness. There is a long history of "explaining" one type of phenomenon simply by its correlation with another. As David Hume (1711–1776) points out in his *Dialogues Concerning Natural Religion*, the Peripatetics, followers of Aristotle, were in the habit of explaining the causes of phenomena in terms of *faculties* or *occult qualities*. Thus, the fact that bread provides nourishment was attributed to its nutritive faculty, and the fact that senn purged was explained by its purgative faculty. Hume bluntly points out that this subterfuge is nothing but a disguise for ignorance.[15]

Even without knowing *how* the brain produces consciousness, scientists would still have grounds for classifying consciousness as an emergent property of the brain if the relation between the brain and consciousness is very similar to the relations between other phenomena and their emergent properties. For example, solidity is a higher level, emergent property of H_2O molecules when they are in a lattice structure (ice); and liquidity is a higher level emergent property of H_2O molecules when they are rolling around on each other (water). The assertion that consciousness is an emergent property of the brain rests, therefore, on the similarity of this relation and that of phenomena like H_2O molecules and the solidity of ice or the fluidity of water. Assuming that such a similarity exists, a great number of contemporary cognitive scientists believe consciousness is a biological feature of certain organisms in the same sense that other features are including photosynthesis, mitosis, digestion, and reproduction.

This view, however, fails to take into account a fundamental difference between mental states as emergent qualities or functions of the brain and these other properties and features: when observing H_2O, one can normally simultaneously observe its liquidity or solidity; and when observing the liquidity or solidity of H_2O, one can normally simultaneously observe the H_2O itself. Indeed, it is hard to imagine how one could detect either of these features of H_2O without detecting the H_2O itself. Similarly, it is hard to imagine how one might examine photosynthesis without at the same time observing a plant, how one could observe mitosis without observing cells or digestion without observing a digestive tract, and so on. In stark contrast to all other relationships between physical phenomena and their emergent properties and functions, when observing the brain, no mental states are observed; and when observing mental states, the brain remains out of sight.

To make this point more clearly, it is quite possible to imagine a single instrument with a kind of zoom lens with which one could detect the liquidity of water at zero magnification; then by greatly increasing the magnification, one could zoom in on the structure of the individual molecules and atoms that make up the same body of water. In contrast, regardless of the magnification of the instrument with which one observes the brain, mental processes themselves are never witnessed; and regardless of the precision of one's firsthand observation of mental processes, the brain processes with which they are associated are never observed.

A genuine emergent property of the cells of the brain is the brain's semisolid consistency, and that is something that objective, physical science can well comprehend. Likewise, scientists clearly understand the mechanisms by which photosynthesis occurs in plants, mitosis occurs within cells, and digestion takes place within a digestive tract; but they do not understand how the brain produces any state of consciousness. *In other words, if mental phenomena are in fact nothing more than emergent properties and functions of the brain, their relation to the brain is fundamentally unlike every other emergent property and function found in nature.* While it is conceivable to learn a great deal about experienced mental states without knowing anything about the brain, and it is feasible to learn a good deal about the brain without knowing anything about subjective mental states, it is not at all clear how one could learn about H_2O at different temperatures and yet know nothing of liquidity or solidity, learn about photosynthesis without knowing anything about plants, and so on in the cases of mitosis, digestion, and reproduction. While liquidity is *perceived* as a quality of water, mental phenomena are only *imagined* as properties of the brain. Indeed, if one were to study the brain alone, while totally ignoring human behavior and subjective conscious states, one would never learn anything about consciousness or any other mental phenomena.

To raise a counterargument in defense of the emergent status of the mind from matter, one could point out that the fluidity of water is indeed a classic example of an emergent property, but it is a primitive one in comparison to the emergence of simple behavior such as an insect's or a robot's ability to walk. Thus, the real dissimilarity between the emergent status of fluidity in water and the emergent status of consciousness from the brain is that the former is a low-level, or primitive, emergence, while the latter is a high-level, or complex, emergence.

While the behaviors of insects and robots is more complex and perhaps pertinent to the mind/body question than the relation between fluidity and H_2O, the relation between the emergent functions of a robot and the robot itself is analogous in a crucial way to the relation between the emergent property of water and H_2O molecules. In both cases, the fact that the function, or property, is in reality an emergent attribute of the robot or water can be ascertained by a single mode of observation, namely, vision. One can directly observe the behavior of a robot and *with the same mode of observation* observe the elemental components of the robot of which that behavior is an emergent function. The robot may *look* like a person, but it is not a conscious person, for it has no interiority, no subjectivity, no dimension of reality other than its "surfaces." But no matter how closely or precisely you visually observe the brain and its functions, you never directly perceive any subjective mental function; and no matter how closely or precisely you introspectively observe mental functions, you never directly observe any brain function. In short, a robot is a lot more like an H_2O molecule than it is like a conscious human. Moreover, the emergent relationship between a robot's behavior and a robot is analogous to that of fluidity and H_2O molecules, but that relationship is, in some utterly crucial ways, not analogous to that between the brain and consciousness.

In rebuttal of this criticism, one might argue that in order to explain the emergence of mental states from the brain, scientists must identify not only the very specific classes of interactions among components of the brain that give rise to mental states but also the specific ways those mental states affect local levels of the brain. Some neuroscientists believe this reciprocal causality between brain states and mental states is the very core of the scientific answer to the mind/body problem. For example, one can train a monkey to desire a particular stimulus in its field of view, and this desire, or anticipation, changes the very minute patterns of neural responses in specific areas of the brain, which, in turn, influence the behavioral outcome. Certain behavior in nonhuman entities is homologous to what humans do, such as moving and remembering, and we not only *see* the behavior but can analyze its emergence out of a complex system, such as a brain, or we can build it, as in the case of a robot. In this sense, behavior becomes the *phenomenal manifestation* of cognition.

Nowhere in this account, however, is there any empirical evidence that any mental state or process is nothing more than an emergent function of specific brain processes. If we attend to the scientific evidence alone—and try to disengage temporarily from the ubiquitous influence of the metaphysical assumptions of scientific materialism—the scientific knowledge we have of the brain is compatible with both of William James's alternative hypotheses: namely, that brain functions simply *allow* for mental events, or that they *transmit* them. A crucial point in this regard is that *the epiphenomenalist view of the mind provides no more intelligible account of body/mind causality than does a dualist view, which is widely dismissed as "unscientific."* This is not to deny the scientific status of the "emergent function" theory of the mind. That is to say, it is indeed a theory that could be refuted in principle by empirical research, for example, if there turns out to be scientifically compelling evidence of clairvoyance, out-of-body experiences, telekinesis, and so on. But those are precisely the areas of research that are deemed to be taboo (under all circumstances) by the upholders of orthodox scientific materialism.

Given the lack of explanatory power of physicalist interpretations of consciousness, why do virtually all contemporary cognitive scientists continue to regard all mental phenomena as functions or properties of the brain? The answer, as suggested previously, may lie in the simple fact that since the Scientific Revolution, natural scientists have been paying attention chiefly to physical phenomena as opposed to mental phenomena. Only physical phenomena have come to be regarded as real, while other phenomena that are not made up of configurations of matter and energy have come to be regarded as unreal. Thus, if mental phenomena are to be admitted into the real world at all, their existence can be acknowledged only as functions or properties of that which is real, namely, physical phenomena. In the more extreme versions of scientific materialism, consciousness and all mental qualia are not counted as existents at all. Recall James's comment that "[h]abitually and practically we do not *count* these disregarded things as existents at all. . . . they are not even treated as appearances; they are treated as if they were mere waste, equivalent to nothing at all."[16]

The Mind As a Nonentity

Recent advances in cognitive science suggest that no physical substances, states, or properties of the brain will be identifiable as the events and states posited by our commonsense experience. The most hard-core proponents of scientific materialism, calling themselves *eliminative materialists*, declare that if this turns out to be the case, it will show that commonsense experience is radically false, and that mental states as we experience them simply do not exist.[17] Paul Churchland, one of the most prominent advocates of this view, declares that commonsense experience is probably irreducible to, and therefore incommensurable with, neuroscience; and for this reason familiar mental states should be regarded as nonexistent or at most as "false and misleading."[18] For similar reasons, philosopher Daniel Dennett bluntly asserts: "[t]here simply are no qualia at all."[19]

If this principle of denying the validity of one area of experience on the grounds that it is irreducible to another were to be applied within other domains of science, a number of our present scientific theories would have to be abandoned. Perhaps the most well-known example of such irreducibility is found within the field of optics. Here we find one large group of phenomena that can be explained only in terms of the wave theory of light and another which can be explained only with the corpuscular theory. How a single entity, such as a photon or an electron, can be both a wave and a particle remains an enigma in quantum mechanics, but it is pragmatically unwise to discard either theory simply because it is not reducible to the other. If this is true within physics, it seems all the more pertinent when dealing with such disparate types of phenomena as are encountered in commonsense experience and in neuroscience.

In denying the very existence of mental states as they are experienced firsthand, eliminative materialists attempt to override experience on purely dogmatic grounds. As noted previously, dogmatists commonly hold to their views even in the face of the most obvious contrary evidence, and their intransigence may grow all the more zealous when obstacles are met. The denial of qualia on grounds that they fail to conform to the principle of objectivism is a clear instance of such irrational adherence to an ideology. The strategy eliminative materialists commonly adopt is to denigrate firsthand experience of our own mental processes with the label "folk psychology." They declare that a fundamental limitation of folk psychological accounts of such mental processes as deliberation, motivation, intention, decision, reason, and desire is that they are inevitably involved in certain stereotyped culture-based outlooks. Some have even suggested that explanations of human conduct by reference to beliefs and desires is comparable to the prescientific explanation of heat by reference to the caloric theory.[20] Moreover, many who take this view assert that folk psychology—which fashions speculative, idealized, culturally contrived accounts of our mental processes in terms of beliefs, desires, and the like—has changed little over the past millennium or more. Thus, they conclude that this subjective approach to understanding the mind is irrevocably flawed and unreliable.[21]

The hypothesis of the existence of a speculative, idealized, and culturally contrived account of any kind of experience that hardly changes from one millennium to another is extremely dubious. If we accept this notion, are we also to believe that folk psychology is not only unchanged through Western history but that it is common to diverse cultures throughout the world? There may indeed be a level of human perception and rationality that is primary, in the sense that humans throughout history, in diverse cultures, experience and understand the world in common. Such primary experiences and judgments are to be contrasted with secondary theories, which we learn and which vary profoundly from one era and society to another.[22] If folk psychology belongs in the former category, it would not be speculative, idealized, or culturally contrived. On the other hand, if it, like the caloric theory, is a secondary theory, it can hardly be said to be an invariant through time or crossculturally.

Eliminative materialists further argue that folk psychological accounts of mental phenomena are undermined by the fact that the actual processes underlying them are in the brain, permanently hidden from our firsthand experience. Thus, unless neuroscience enlightens us, they must remain a mystery to us. All folk psychological models of mental processes, they claim, will necessarily be inferior to any rival model based on better information as to how our brains function or to any model built up in terms of an ideally rational, artificial intelligence explanation based on the methods of some pure deductive or inductive logic.[23]

Proponents of this view place their hopes in future advances in the brain sciences, when terms such as *belief, desire, hope*, and *intention* may be gradually replaced by neurophysiological terms. However, at present scientists can only guess at what that new language may turn out to be. Thus, the alternative to subjective folk psychology is objective brain science and computer science, but it is not these sciences of the past or present that have the cure for the inadequacies of past and present folk psychology. Rather, these scientific materialists look to *future* science that will provide both a new language to replace our familiar cognitive, emotive, and appetitive terminology and new, objective knowledge to replace our old, subjective, folk psychology. In this way the taboo against subjectivity is maintained at all costs; and instead of relying on experience, eliminative materialists encourage us to place our faith in scientific knowledge that does not presently exist but may one day emerge.

For fifty years, behaviorism, with its claim that subjective mental states are nonexistent, sought to override common sense; but in the end, it was behaviorism that was discarded, not common sense. Eliminative materialists, in turn, designate reports of subjective experience as folk psychology, and treat that as if it were just one more secondary theory that we acquire through study. In fact, many elements of folk psychology may constitute our primary experience; while the speculative, idealized, culturally contrived principles of scientific materialism constitute a secondary theory that is unique to modern Western society and other civilizations recently influenced by the West.

Those who deny the very existence of subjectively experienced mental states proclaim that their position is comparable to Galileo's insistence on a heliocentric view of the solar system. The glaring dissimilarity between these two cases is that Galileo had fresh, empirical knowledge in support of his effort to overthrow a tradition-bound, rationalistic ideology. Eliminative materialists, on the other hand, have no new empirical knowledge to explain the nature or origins of conscious states; rather they dismiss subjective experience simply on the grounds that it fails to conform to their rationalistic ideology. In this way, they resemble Galileo's opponents far more than Galileo himself.

The Causal Agency of the Mind

The question of the causal efficacy of the mind has plagued scientific materialism since the time of Descartes, and neither dualists nor physicalists have provided a compelling solution to this problem. If the mind is so different from matter that it has no mass, no shape, and no location in space, how is it possible for it to have any causal influence on the body or anything else? If the mind can in fact be equated with certain processes or functions of the brain, all causal efficacy can be attributed to those physical events. But what of the causal agency of qualia? Do our subjectively experienced perceptions, thoughts, intentions, desires, feelings, beliefs, and so on not influence our behavior? If qualia were simply nonexistent, none of us, including neuroscientists, would be able to perceive, understand, or actively engage with the world about us. Nothing could be more obvious than the fact that perceptual appearances and mental events of joy and sorrow, hopes and fears, thoughts and mental imagery are key elements in the world of our experience.

If these qualia have no causal influence in the natural world, what good are they from an evolutionary perspective? If they have no real function in human evolution, how did they happen to arise in the first place? To survive and procreate, an organism simply has to detect its environment in appropriate ways, compute appropriate responses, then mechanically carry out those responses. All these tasks might well be performed without consciousness and without qualia, so the experiential fact that we are conscious of a world of human experience seems—from an evolutionary perspective—quite useless.[24]

The major factor preventing scientific materialists from admitting the existence of qualia is the sense that they would have to be given a causal role with respect to the physical world and especially the brain. And taking this step appears to take one down a slippery slope to an antiscientific belief in spirits, demons, and fairies. Remarkably, the threat of the preternatural, despite its metaphysical exorcism from the realm of science in the seventeenth century, still seems to haunt scientific and philosophical thinking today. Even recently, scientific materialists have continued to attack accounts of demonic possession and the efficacy of exorcism. Such beliefs, they argue, are incompatible with our present understanding of physics and chemistry. In fact, they claim, their falsity is directly discoverable without considerations having to do with chemistry, physics, or biology. Suitably controlled experiments would quickly demonstrate that exorcism does not affect devil possession–type behavior and that certain other therapies do.[25]

Recall that in 1665, Thomas Sprat had already denied the possibility of demonic possession and claimed that the nonexistence of demons and fairies had already been demonstrated by *experiments*, without citing which experiments those might be. Now, more than three centuries later, contemporary scientific materialists claim that *if experiments were run*, they would demonstrate the causal inefficacy of exorcism. They acknowledge that even though some cases of exorcism might *appear* to be efficacious, this is only because they involve certain features of therapies authorized by scientific materialism. Scientific proof of the nonexistence of demons may require some ingenuity, they claim, but it poses no problem *in principle*. In short, since the seventeenth century, proponents of a mechanical view of nature have argued that such notions are simply irrational, and scientific validation of their own beliefs is easily available, *but for some unexplained reason has not been demonstrated.*

Thus, in the apparent absence of hard, scientific evidence one way or the other, scientific materialists assure us that any assertion of the efficacy of exorcism is "just false" and proving this scientifically poses no problem "in principle." The underlying assumption of any such proof seems to be that a mechanical explanation of any apparent efficacy of exorcism must rationally be accepted over against any other explanation that affirms the existence of a preternatural realm. In short, it is not sufficient to account for the empirical facts; they must be accounted for in terms of mechanical, physicalist view of nature.

Scientific materialists deny the existence and causal efficacy of nonphysical mental events on the grounds that (1) no compelling explanation for any causal interaction between a nonphysical mind and the body has been devised, and (2) the principle of the conservation of mass/energy would seem to preclude such a nonphysical intervention into the physical world. Thus, since no causal mechanism between the two has been determined, such causality is denied, even though this seems to fly in the face of firsthand experience. Paradoxically, scientific materialists insist that all mental events, including consciousness itself, are produced solely by the brain, despite the fact that no causal mechanism has been identified in the brain that explains how consciousness arises. Since physicalists have been no more successful at explaining the causal interaction between the mind and body than the dualists, the only serious objection to considering the causal efficacy of a nonphysical mind in the physical world is the conservation principle.

This issue must be reconsidered within the context of quantum mechanics, which does not necessarily support the closure principle. According to quantum theory, the so-called energy-time uncertainty principle does allow for short violations of energy conservation. Thus, it is possible in principle for a nonphysical mind to engage with so-called matter. But before exploring this hypothesis in detail, it may be important to re-examine what the term matter even means in contemporary physics. Nobel laureate Steven Weinberg comments: "[i]n the physicist's recipe for the world, the list of ingredients no longer includes particles. Matter thus loses its central role in physics. All that is left are principles of symmetry."[26] These principles are patterns, or relationships, the very existence of which, as purely objective phenomena independent of the mind that observes or conceives of them, is questionable.

Most physicists agree there are no physical causes for individual quantum events, and they conclude from this that individual quantum events are fundamentally random; that is, there are no preceding causes that determine them. But the absence of physical causes does not preclude the possibility of nonphysical causes. Physicists have found that even if there were local causes for specific quantum effects, they must be *physically* undetectable *in principle;* but that doesn't necessarily mean they are *absolutely* undetectable in principle. On the quantum level, unknown causal agencies may be posited without contravening the conservation principle if, for any given system of measurement, (1) one does not (or cannot) specify the complete, exact initial conditions of the system to be measured and (2) one allows for nonlocal influences. As for the first point, the uncertainty principle, together with the physical impossibility of absolutely isolating any finite system of measurement make it impossible to determine the complete initial conditions of any system. As for the second point, recent research by Anton Zeilinger and other physicists indicates that there are strong grounds for asserting the reality of nonlocal interactions. In short, even though physicists know there are no local causes for quantum events, there could be nonlocal ones.

Given the startling conclusions of quantum mechanics, it is remarkable that its insights have had so little impact on science as a whole. Although quantum effects vanish statistically in the macro-world—which is to say, there are intelligible accounts of observed phenomena without reference to the uncertainties of quantum mechanics—the ontological questions raised in quantum theory do not disappear. The founders of quantum mechanics thought their discoveries would revolutionize all of science, but the Second World War intervened, and the theoretical implications were superseded by the practical applications of modern physics. Following the war, the trend of scientific research continued to be materialistic and pragmatic in orientation, so the theoretical implications of quantum mechanics were largely quarantined apart from other scientific disciplines. Fortunately, public and scientific fascination with the implications of quantum mechanics now appears to be on the rebound.

Why do so many scientists and philosophers continue to assume that scientific materialism constitutes the most promising set of hypotheses for exploring consciousness? Presumably because it has been enormously successful for understanding a wide array of objective phenomena. But it has left us in the dark concerning the mind/body problem, the nature of consciousness, and subjective mental phenomena in general. Contemporary philosophy of mind is simply at a stalemate when it comes to the question of consciousness, with some philosophers regarding the nature of qualia and consciousness as a "hard problem" and others not.

While scientific materialists generally acknowledge their ignorance of the origins, nature, and function of consciousness, they place their faith in future discoveries in the neurosciences to answer these questions. While the neurosciences have effective methods for exploring the brain, when it comes to consciousness, they have not come upon empirically verifiable surprises that force scientists to make substantial revisions in their basic description of reality; and they have not demonstrated that they can achieve the goal of challenging, and perhaps transcending, the fundamental assumptions under which they operate. In his book *The End of Science: Facing the Limits of Knowledge in the Twilight of the Scientific Age*, John Horgan coins a term for disciplines that suffer from such shortcomings: "ironic science." Such a science, he says, "cannot achieve its goal of transcending the truth we already have. And it certainly cannot give us—in fact, it protects us from—*The Answer*, a truth so potent that it quenches our curiosity once and for all time."[27] If we are to find the truth concerning the origins, nature, and function of consciousness, it seems we must use other modes of inquiry, which will violate the taboos of scientific materialism.

CONFUSING SCIENTIFIC
MATERIALISM WITH SCIENCE

After four centuries of advances in scientific knowledge, more than a century of psychological research, and roughly a half century of progress in the neurosciences, even most advocates of scientism acknowledge that science has yet to give any intelligible account of the nature of consciousness. Nevertheless, the extent of our ignorance concerning consciousness is often overlooked. This ignorance is like a retinal blind spot in the scientific vision of the world, of which modern society seems largely unaware. In most books and articles on cosmogony, evolution, embryology, and psychology, consciousness is hardly mentioned; and when it is addressed, it tends to be presented not in terms of experiential qualia but in terms of brain functions and computer systems.

Under the doctrinal influence of scientific materialism, the public has been led to believe that scientists know things about the mind of which they are in fact ignorant and to believe that ordinary human subjects do not know things that they do in fact know perfectly well. A major tendency of scientific materialism has been to describe machines and other unconscious phenomena in anthropomorphic, cognitive terms. The same terms, adjusted to their application to machines, are then reapplied to human minds, giving the impression that minds and machines are essentially alike. Thus, a kind of "neuromythology" is fabricated that simultaneously obscures the actual nature of both machines and minds.[1]

The broader, ubiquitous problem is one of confusing the metaphysical assumptions of scientific materialism with the empirical knowledge of science. This tendency commonly appears in journalistic accounts of science, science textbooks, and learned philosophical writings. To illustrate this point, I offer three case studies: a recent cover article in *Time* magazine on the nature of the mind; a commonly used textbook on cognitive psychology; and a recent, influential philosophical work on the nature of the mind.

Journalistic Confusion

The cover story of *Time* magazine on July 17, 1995 is an article entitled "Glimpses of the Mind: What Is Consciousness? Memory? Emotion? Science unravels the Best-Kept Secrets of the Human Brain," by Michael D. Lemonick. This article in various ways unconsciously substitutes metaphysical assumptions of scientific materialism for genuine scientific discoveries, as follows.

The first page of the article shows a computer-generated image of the brain based on a positron-emission tomography (PET) scan, with a caption describing this as an image "of a sad thought." In reality, it shows an area of the brain in which there is increased metabolism in the most active cells while a sad thought is present. Apart from a human subject's report of experiencing a sad thought, no such association could have been made with this portion of the brain. Moreover, the equivalence of a sad thought with specific neuronal activity is a hypothesis that has in no way been established scientifically. The author simply assumes the validity of the metaphysical assumption of the identity of specific mental states and specific brain states, without informing the reader of his choice or the reasons for it.

In a similar vein, a heading over two consecutive pages reads: "[a] memory is nothing more than a few thousand brain cells firing in a particular, established pattern." This statement is merely an affirmation of the principles of reductionism and physicalism and like scientific materialism as a whole, completely ignores the qualia associated with memories. Continuing with his uncritical equation of brain and mind events, the author writes:

> [p]owerful technologies such as magnetic resonance imaging (MRI) and positron-emission tomography (PET) have also provided a window on the human brain, letting scientists watch a thought taking place, see the red glow of fear erupting from the structure known as the amygdala, or note the telltale firing of neurons as a long-buried memory is reconstructed.

While Lemonick reports Francis Crick's speculation that the general principles of a physiological reduction of visual qualia might be within our grasp *before the end of the twentieth century*, he apparently fails to recognize this as a tacit admission that such scientific knowledge does not now exist. This fact is also obscured by psychiatrist Larry Squire's reductionist claim that the combination of all the patterns of neuronal connections "gives you a complete perception." The qualia of perception, as usual, are simply ignored. Lemonick then goes on to quote neuroscientist Rodolfo Llinás's assertion that "light is nothing but electromagnetic radiation." If the term "light" refers here to physical light that travels through space at 186,000 miles per second, this statement conveys nothing new. On the other hand, if it is meant to refer to the *qualia* of light, this assertion is nothing more than an affirmation of reductionism, with no empirical validation. Llinás goes on to assert that sound is the relation between external vibrations and the brain, and again he ignores the qualia of sound altogether.

Turning to the topic of consciousness per se, Lemonick writes: "[i]t turns out that the phenomenon of mind, of consciousness, is much more complex, though also more amenable to scientific investigation, than anyone expected." To illustrate his point, he cites one scientist's speculation after another, without presenting any concrete scientific evidence to support these conjectures. For example, neurologist Antonio Damasio is quoted as speculating that, contrary to common sense, "consciousness may be nothing more than an evanescent by-product of more mundane, wholly physical processes." Lemonick also reports the speculation of Francis Crick and Christof Koch that consciousness "is somehow a by-product of the simultaneous, high-frequency firing of neurons in different parts of the brain." While Crick himself acknowledges this is not a scientific discovery but a highly speculative concept, Lemonick assures the reader that it is to be taken seriously, for such theories of consciousness "are at least firmly rooted in biology." Thus, these claims are backed, not by empirical evidence or compelling reasoning, but by sheer authority.

In a similarly facile manner, Lemonick dispenses with the problem of human identity with the following declaration:

After more than a century of looking for it, brain researchers have long since concluded that there is no conceivable place for such a self [i.e., some entity deep inside the brain that corresponds to the self] to be located in the physical brain, and that it simply doesn't exist.

If modern neuroscience does not understand the nature of sensory qualia, and if it is ignorant of the nature of consciousness, this assertion concerning the nature of the self can hardly be anything more than sheer speculation. Moreover, the grounds of this conclusion—that it cannot be located anywhere in the brain—is simply an affirmation of the principle of physicalism. Lemonick raises a weak challenge to this principle in his concluding comment: "[i]t may be that scientists will eventually have to acknowledge the existence of something beyond their ken—something that might be described as the soul." This comment is similar to the God-of-the-gaps approach to affirming the existence of the Deity: attribute to God whatever cannot be explained by science. But this strategy offers no genuine understanding of God, the soul, or anything else; it is therefore equally unsatisfying to religious believers and scientists alike. In short, while scientists and journalists are free to speculate on the nature of the mind, this article misleads its readers by failing to distinguish between scientists' metaphysical speculations and genuine scientific knowledge.

Pedagogical Confusion

The conflation of metaphysical speculation with scientific knowledge is also very prevalent in textbooks on the cognitive sciences. The rise of cognitive psychology around the middle of the twentieth century has often been heralded as a return to the study of the mind, which was deliberately ignored by American psychologists during the fifty-year reign of behaviorism. However, as the following critique of John Anderson's textbook *Cognitive Psychology and Its Implications* will demonstrate, that reputation is only partially deserved.

Anderson represents the dominant viewpoint in cognitive psychology by taking an "information processing" approach to the study of the mind. In this context, information is thought to be represented in terms of continuously varying electrochemical activity of neurons.[2] This neuronal activity is discontinuous, episodic, and often quantal. Thus, information may be carried by bursts or by a variety of patterns, with more than one neurotransmitter being released from the same terminal.

To take a specific example of such information processing, Anderson asserts that visual perception begins with energy from the external environment; and receptors, such as those on the retina, transform this energy into neural information (83). In this process, light is converted into neural energy by a photochemical process, and low-level cells in the visual system "detect simple patterns of spots of light and darkness in the visual field" (19).

Physicists understand light as consisting of quantized electromagnetic energy of various frequencies, traveling through space at 186,000 miles per second. Such energy may indeed be converted into neural energy, but Anderson gives no account of any process by which such energy becomes transformed into the *qualia* of lightness and darkness in a visual field. Nonconscious, low-level cells may *react* to impulses of energy, but where is the evidence that they *detect* experiential patterns of lightness or darkness in a perceived visual field? Such an experiential visual field does not travel through space at the speed of light, but we are left with the questions: where do visual qualia exist, what is their nature, and what are the sufficient causes for their occurrence? While neuroscience sheds a great deal of light on the biological basis of vision, it is not clear that it addresses these issues or that it has the experimental means to investigate them. This omission is concealed by the *objectivist* use of the cognitive term *detect* with respect to an *experiential* visual field. The mind/body problem has not been solved but rather has been camouflaged by this use of terminology.

This point has direct bearing on the endeavor to design computer models to simulate the information processing that takes place in the visual system. According to Anderson, the goal of such research is "to get these computer programs to *see* in a visual scene what a human sees" (37). Humans experientially see visual qualia in dependence on unconscious interactions between the environment and the nervous system; and neuroscientists now generally acknowledge that there is no one-to-one correspondence between the frequencies and intensities of objective light and the qualia that make up our visual world. If Anderson is suggesting that the computer programs he has in mind are designed to *consciously experience* the visual world perceived by humans, he must identify which components of these programs produce consciousness and explain the mechanism for this occurrence. If, on the other hand, such computers are designed to react *nonconsciously* to objective light, then there must be a one-to-one correspondence between their reactions and the objective features of the incoming electromagnetic energy. But in this case, it would follow that those mechanical devices are guaranteed *not* to see in a visual scene what a human sees. Indeed, we have no compelling reason to believe that they experientially *see* anything, nor are there any generally accepted criteria for judging what would count as evidence that mechanical devices *do* actually see.

In Anderson's presentation of the neural basis of cognition, cognitive terms are uniformly objectified. He asserts, for example, that "[c]ognition is achieved by patterns of neural activation in large sets of neurons," and "resides in patterns of the primitive elements of computers" (18,24). While he declares that "the brain encodes cognition in neural patterns" (24), he acknowledges that no one knows *how* this occurs. And he offers no justification or explanation for asserting that brain cells experientially *detect*, rather than merely *electrochemically react to*, visually related physical stimuli.

Anderson's account postulates not only unconscious cognitive processes but ones of which *we cannot be aware*. This hypothesis is based on the assumption that there is no essential or necessary connection between computation and consciousness. Anderson extends this principle beyond cognition to emotions when he writes that "computer systems... have been shown to be capable of... displaying frustration" and this "feeling of frustration" occurs in "large patterns of bit changes" (24). He offers no evidence for the presence of this emotion, nor does he demonstrate the manner in which feelings can become embedded in patterns of bit changes. Thus, this is one of the most flagrant examples of neuromythology, based not on scientific evidence, but on the unquestioned assumptions of scientific materialism.

There is a good deal of mystery surrounding the questions of how the subjective experiences of cognition and even emotions are supposed to be achieved by the components of the brain and the computer. Anderson deals with this mystery by answering: "[i]t does not appear that there is anything magical about human intelligence or anything that is incapable of being modeled on a computer" (3), but he offers no justification for divorcing cognitive and affective terms from conscious experience and imputing them upon nonconscious, material objects and processes. Contrary to his claim, there is, in fact, a facet of human intelligence that does not appear to have been modeled on a computer—and that is consciousness. Without addressing this issue, we are poorly equipped to answer the question: are patterns of neural activation merely *necessary* for conscious cognition to be achieved, or are they *sufficient?* Anderson's account not only sheds no light on this question but obscures that there is any such problem at all.

Similar confusion occurs in this textbook's account of mental imagery. While cognitive psychology, unlike behaviorism, commonly acknowledges the existence of such mental qualia, Anderson explains it away with the comment that "when subjects are scanning a mental array, they are scanning a representation that is analogous to [a] physical array" (96). Thus, a mental image is "an abstract analog of a spatial structure" (98), and certain data might seem to indicate that subjects rotate mental objects in a three-dimensional space "within their heads" (93). Anderson hastens to add that "subjects are not actually rotating an object in their heads" (93), but he does not explain where mental objects *are* rotated. If not in the head, it is even less likely that such objects exist outside of the head; and this raises the question: where, if anywhere, do they exist? As usual, this mechanistic account of the mind fails to illuminate the actual nature and origins of qualia of any kind. While scientific theories are characteristically based on and tested by means of empirical evidence, metaphysical dogmas are based on unquestioned assumptions and are immune to empirical evidence. Anderson's textbook account of cognition and consciousness evidently falls into the category of metaphysical speculation, while falsely posing as scientific knowledge.

Philosophical Confusion

An Empirical Challenge to the Taboo against Subjectivity

Many modern philosophical accounts of the mind are sophisticated expressions of scientific materialism. However, John Searle's influential work *The Rediscovery of the Mind* is more complex, for it raises fundamental experiential objections to many of the contemporary rationalistic accounts of the mind/body problem. In a striking departure from the more orthodox views of the mind according to scientific materialism, Searle points out the disastrous effects resulting from the failure of modern philosophers and psychologists to come to terms with the subjectivity of consciousness. Indeed, he declares that much of the bankruptcy of most work in the philosophy of mind and a great deal of the sterility of academic psychology over the past fifty years have come from a persistent failure to recognize and come to terms with the fact that the mind is an irreducibly first-person phenomenon.[3] Searle goes further in pinpointing the rationale for the perpetuation of that approach.

Acceptance of the current views is motivated not so much by an independent conviction of their truth as by a terror of what are apparently the only alternatives. That is, the choice we are tacitly presented with is between a "scientific" approach, as represented by one or another of the current versions of "materialism," and an "antiscientific" approach, as represented by Cartesianism or some other traditional religious conception of the mind. (3–4)

The disaster of this taboo of subjectivity stems from the strategy of describing the world as completely objective, leaving out subjectivity altogether, which has been a central premise of scientific materialism in general and modern cognitive science in particular. This strategy, he points out, makes it impossible to describe consciousness, because it becomes literally impossible to acknowledge the subjectivity of consciousness. Thus, for all its purported rejection of the Cartesian framework, cognitive science has maintained the absolute dichotomy of conscious, subjective mental processes—which are not regarded as a proper subject of scientific investigation—and objective neurological and behavioral processes that are regarded as the genuine subject matter of science.

In recent decades, Searle points out, because of their inability to explain consciousness, cognitive scientists have made systematic efforts to dissociate consciousness from intentionality. This "objective" treatment of intentionality implies that the subjective features of consciousness are irrelevant to intentionality. In short, "[m]ore than anything else, it is the neglect of consciousness that accounts for so much barrenness and sterility in psychology, the philosophy of mind, and cognitive science" (227).

In the second chapter of *The Rediscovery of the Mind*, Searle sums up and concisely demonstrates the philosophical inadequacies and violation of experience in the recent, prominent theories of the mind, including behaviorism, type-identity theories, token-token identity theories, black box functionalism, strong artificial intelligence, and eliminative materialism. After effectively refuting these materialistic theories, he sets forth his own views concerning the nature and origins of mental phenomena.

He begins by asserting that when examining the existence of mental states as mental states, the correlated behavior is neither necessary nor sufficient for their existence. That is, a mental state may arise and pass without the occurrence of any correlated, externally observable behavior; and the event of a specific type of behavior has no necessary, one-to-one relation with any specific type of mental state, or intentionality. Mental states are ontologically irreducible, they exist only as subjective, first-person phenomena, and they almost always have a content. While it is perfectly legitimate, he maintains, to ask how unconscious bits of matter in the brain produce consciousness, the individual neurons (or synapses or receptors) in the brain are not themselves conscious.

Central to Searle's conception of the mind is his "connection principle," which asserts that "[t]he notion of an unconscious mental state implies accessibility to consciousness. We have no notion of the qualia of the unconscious except as that which is potentially conscious."[4] The class of deep unconscious, mental, intentional phenomena that are not only unconscious but that are in principle inaccessible to consciousness simply does not exist. "Not only is there no evidence for their existence, but the postulation of their existence violates a logical constraint on the notion of intentionality" (173). While consciousness comes in a variety of modalities—including perception, emotion, thought, pains, and so on—talk of the unconscious mind is merely talk of the causal capacities of the brain to cause conscious states and conscious behavior. In short, Searle's basic position is that the study of the mind is the study of consciousness. Moreover, "the specifically mental aspects of the mind can be specified, studied, and understood without knowing how the brain works. Even if you are a materialist, you do not have to study the brain to study the mind" (44).

A Rationalist Retreat to Scientific Materialism

Despite Searle's revolutionary suggestions for a truly empirical study of the mind as a first-person phenomenon, his own theory of the nature and origins of the mind retreats to the creed of orthodox scientific materialism. In a familiar refrain, he echoes the prevailing view that mental phenomena are caused by neurophysiological processes in the brain and are themselves features of the brain (1); this knowledge, he says, has been available to all educated people for at least a century, or since the beginning of the scientific study of the brain. Searle adopts the conventional view that consciousness is an emergent property of the brain in the sense that liquidity is a property of systems of water molecules; and he rejects the Cartesian mind/matter dualism by suggesting that consciousness is both *physical and mental*. The obvious failure of this idea, however, is that he provides no intelligible account of what it is about the brain that enables it to possess mental properties. Thus, his designation of mental phenomena as being both physical and mental appears to be nothing more than a shift in terminology, without elucidating anything new about consciousness.

As noted earlier, contemporary brain science has no objective means of detecting the presence of consciousness, whether in primitive organisms such as a hydra, in a developing human fetus, in an adult human, or in a computer; nor does it have any compelling explanation of the manner in which consciousness is produced. In light of this uncontested fact, what are the grounds of Searle's assertion that knowledge of the nature and origins of the mind has been widely available since the rise of modern neuroscience? Is this really anything more than an appeal to the authority of scientific materialism? Searle acknowledges that we are at present very far from having an adequate theory of the neurophysiology of consciousness, and that "[w]e would . . . need a much richer neurobiological theory of consciousness than anything we can now imagine to suppose that we could isolate necessary conditions of consciousness" (91,76–77). In short, neuroscientists simply do not know how consciousness is produced; but Searle rests in the firm conviction that its production occurs in human brains in virtue of specific, "though largely unknown," features of the brain (89, 57).

Having candidly acknowledged that contemporary neurophysiology does not know how consciousness is produced, Searle simply affirms his status as a scientific materialist by claiming that "[c]ausally we know that brain processes are sufficient for any mental state" and that "consciousness is entirely caused by the behavior of lower-level biological phenomena" (23,92). These two statements necessarily imply a thorough knowledge of *all* the specific causes necessary for the production of consciousness, but contemporary neurophysiology certainly does not possess such knowledge. Searle's justification for this violation of reason seems to lie in his faith in the future of the neurosciences: "[i]f we had an adequate science of the brain, an account of the brain that would give causal explanations of consciousness in all its forms and varieties, and if we overcame our conceptual mistakes, no mind/body problem would remain" (100).

However, in the real world of the present, there is no such "adequate science of the brain"; there is no "account of the brain that would give causal explanations of consciousness in all its forms and varieties"; modern scientific and philosophical theories of the mind are riddled with "conceptual mistakes"; and the "mind/body problem" is still very much with us. Thus, Searle's assertion that "[t]he existence of consciousness can be explained by the causal interactions between elements of the brain at the micro level" (112) is an expression of faith in *scientific knowledge that does not exist*.

As noted earlier, Searle insightfully points out that the strategy of scientific materialism—to describe the world as completely objective, leaving out subjectivity altogether—makes it impossible to describe consciousness because of the impossibility of acknowledging its innate subjectivity. But most contemporary neuroscientists are absolutely committed to this strategy, a point that Searle apparently overlooks. Thus, when raising the question of whether fleas, grasshoppers, crabs, or snails are conscious, he suggests (74) that such questions can reasonably be left to neurophysiologists—who have no scientific criteria for identifying consciousness and who find it impossible to describe! Searle's dogmatic faith in scientific materialism does not stop there. Upon posing the question of the spatial location and dimensions of conscious experience, he removes this issue from the arena of first-person conscious experience and describes it as "an extremely tricky neurophysiological question" (105). This is all the more surprising in light of his previously mentioned assertions that the ontology of the mental is an irreducibly first-person ontology and that mental states exist only as subjective, first-person phenomena.

To take a specific case, Searle points out that common sense indicates that our physical pains are located in the physical space within our bodies— that a pain in the foot, for example, is literally inside the area within the foot. This is the subjective, first-person account of the phenomenon of an experienced pain in the foot. But Searle overrides this account by declaring that the brain forms a body image, which exists in the brain, and that all bodily sensations are parts of this body image. Thus, contrary to subjective, first-person experience, the pain in the foot is actually in the physical space of the brain (63). While first-person experience is certainly fallible, as in the case of pain felt in an amputated limb, this does not imply that all subjective experiences of pleasure and pain are misleading.

Searle's assertion that the location of the body image is literally inside the brain must be based solely on the fact that the brain produces that image—an assertion that ignores all the other internal and external influences that contribute to the occurrence of that experience of pain. If this is true, would it not be equally plausible to suggest that the other experienced attributes of the foot—including its color, smell, and texture—exist in the physical space of the brain and not in the foot? If it is justifiable to discard first-person, subjective experience of a pain in the foot, why not discard it for these other factors that must be associated with our body image? If so, why should we stop with the foot? Why not lodge the rest of the body as it is subjectively experienced into the physical space of the brain? .

Following this line of thought, the brain should produce not only a body image but images of the rest of the physical world as well. In this case, all the colors, sounds, smells, tastes, and textures that are perceived as (the Cartesian secondary) attributes of the physical world according to first-person subjective experience actually exist only within the physical space of the brain. This can only mean that subjective experience is totally misleading when it comes to perception and that we must rely entirely on neurophysiology to determine the location of such experiences. But we are now left with the same problem as before: neurophysiology has no objective way of detecting or explaining subjective states of consciousness. No inspection of the brain has ever revealed a "body image" inside the skull, and it is questionable whether it will ever be found. Thus, Searle's assertion that the pain in the foot actually exists in the brain is reminiscent of Groucho Marx's challenge: "Who are you going to believe—me or your very own eyes?"

Self-consciousness and Introspection

Following Searle's assertion of the irreducibility of conscious mental states that exist solely as first-person, subjective phenomena, he poses the question: how are such phenomena to be studied scientifically? In response, Searle initially takes the empirical approach of suggesting that we let our research methods dictate the subject matter, rather than the converse: "[b]ecause mental phenomena are essentially connected with consciousness, and because consciousness is essentially subjective, it follows that the ontology of the mental is essentially a first-person ontology . . . The consequence of this . . . is that the first-person point of view is primary" (20). Searle rightly cautions that it is immensely difficult to study mental phenomena and the only guide for methodology is the universal one, namely, to use anything that works.

To carry through with this pragmatic dictum, however, is difficult when it comes to studying subjective consciousness scientifically. Adhering to the principle of objectivity, science deals with "empirical facts" that are testable by "empirical methods," and this traditionally entails testability by third-person means. But this methodology, Searle insists, entails the false assumption

> that all empirical facts, in the ontological sense of being facts in the world, are equally accessible epistemically to all competent observers. We know independently that this is false. There are lots of empirical facts that are not equally accessible to all observers. (72)

This very assumption, of course, makes the scientific study of all uniquely first-person accounts of mental phenomena highly problematic. Following the dictates of scientific materialism, with one fell swoop it removes all subjective events from the realm of empirical facts.

The obvious implication of Searle's view of consciousness is that mental phenomena must be studied primarily from a first-person perspective. In this regard, he acknowledges the existence of "self-consciousness," which he describes as "an extremely sophisticated form of sensibility . . . [that] is probably possessed only by humans and perhaps a few other species" (143). Such consciousness is "directed at states of consciousness of the agent himself and not at his public persona" (142) and entails awareness of one's mental and physical behavior. Searle goes on to make the experiential claim that just as we can shift our attention from the objects at the center of consciousness to those at the periphery, we can also shift our attention from the *object* of conscious experience to the *experience* itself. "In any conscious state," he asserts, "we can shift our attention to the state itself. I can focus my attention, for example, not on the scene in front of me but on the experience of my seeing this very scene" (143).

Thus, Searle apparently opens the door to the possibility of scientific introspection of mental phenomena; but, in an abrupt withdrawal from experience back into the dogma of scientific materialism, he utterly rejects this possibility:

> if by "introspection" we mean a special capacity, just like vision only less colorful, that we have to *spect intro*, then it seems to me there is no such capacity. There could not be, because the model of specting intro requires a distinction between the object spected and the specting of it, and we cannot make this distinction for conscious states. (144)

The reason for this, he asserts, is that while the model of vision works on the presupposition that there is a distinction between the things seen and the seeing of them, for "introspection" there is simply no way to make this separation. "Any introspection I have of my own conscious state is itself that conscious state . . . the standard model of observation simply doesn't work for conscious subjectivity" (97). Moreover, just as the metaphor of introspection breaks down when the only thing observed is the observing itself, so does the metaphor of a private inner space break down because of the impossibility of making the necessary distinctions between the three elements of oneself, the act of oneself entering such an inner space, and the space into which one might enter. In the conclusion of his refutation of introspection Searle writes:

> [w]e might summarize these points by saying that our modern model of reality and of the relation between reality and observation cannot accommodate the phenomenon of subjectivity. The model is one of objective (in the epistemic sense) observers observing an objectively (in the ontological sense) existing reality. But there is no way on that model to observe the act of observing itself. For the act of observing is the subjective (ontological sense) access to objective reality (99).

Searle's rejection of introspection is strikingly at variance with the rest of his presentation of consciousness and his endorsement of "self-consciousness." While the model of vision endorsed in scientific materialism is indeed based on the presupposition of the absolute Cartesian distinction between subject and object, this model is not what allows for the experienced reality of sight. First-person *experience* is our basis for asserting the reality of vision and is also the basis of Searle's assertion of the reality of self-consciousness. What he apparently fails to recognize is that the same "modern" model of reality (namely, scientific materialism) that rejects the possibility of introspection equally leaves no room for his experience of self-consciousness. This model of reality assumes a duality between observed phenomena, which are absolutely objective, and the observations of them, which are absolutely subjective. Thus, Searle's own arguments for rejecting introspection on dogmatic grounds equally refute his own experience of self-consciousness, which bears all the earmarks of introspection.

As noted earlier, Searle recognizes the debilitating failure of scientific materialism to accommodate the phenomenon of subjectivity. Yet now, when faced with its inability to account for the experience of the first-person observation of mental phenomena, he rejects not the faulty dogma but the subjective experience that is incompatible with it! This position is all the more internally problematic in light of Searle's assertion that in the case of consciousness, there is no appearance-reality distinction, for the reality of consciousness is the appearance (122). The statement that consciousness is an appearance of any kind makes sense only if consciousness appears, but according to the dogma with which he refutes introspection, only the objective world, and not subjective consciousness, ever appears.

To bolster his argument, Searle points out that while it is easy to describe visually perceived objects on a table in front of one, it is difficult to describe, separately and in addition, one's conscious experience of those objects (127). But Searle apparently does not consider the alternative of initially describing the *conscious experience of the visual appearances* of those objects, which would then leave the difficult task of describing, separately and in addition, those *objects* apart from the conscious appearances of them. The problem lies not in the inaccessibility of consciousness, as Searle implies, but in the absolute reification of subjective consciousness versus objective reality.

The brunt of Searle's rationalistic argument is that for something to be observed, it must be separate and independent from the consciousness of it; and since conscious states do not fulfill that criterion, they must be unobservable (whether they seem to be experientially observable or not). Thus, he argues, vision is possible because the objects of vision are separate and independent of sight, and the same holds true for the rest of our sensory experience of physical reality. According to this view, which Searle apparently endorses, all particles, systems, organisms, and so on that make up the real world are "completely objective," in consequence of which they are equally accessible to all competent observers (96).

This endorsement of objectivism is at variance with the fact that all our normal perceptions of "objective" reality are always structured by such "subjective" factors as memories, desires, and expectations. Thus, as Searle acknowledges,

> these features hang together: structuredness, perception as, the aspectual shape of all intentionality, categories, and the aspect of familiarity. Conscious experiences come to us as structured, those structures enable us to perceive things under aspects, but those aspects are constrained by our mastery of a set of categories, and those categories, being familiar, enable us, in varying degrees, to assimilate our experiences, however novel, to the familiar. (136)

These points hold equally true for everyday experience as well as for scientific observation, in which a considerable amount of conceptual training and at times indoctrination is generally required before one achieves the status of being a "competent observer." In light of these many subjective conceptual and perceptual influences on our experience of reality, there is a hollow ring to Searle's statement that everything in the real world is completely objective and equally accessible to all competent observers. In short, upon careful examination of everyday experience and scientific observation, it seems contradictory to assert the possibility of observing independently existing physical phenomena in the objective world while simultaneously asserting the impossibility of observing mental phenomena in the subjective world.

Despite Searle's criticisms of contemporary scientific materialism, with its physicalism and reductionism, he dogmatically insists that modern educated people are compelled to accept this view and not even consider it to be in competition with other, incompatible worldviews (90). Thus, hypotheses affirming the existence of God or an afterlife are not to be taken seriously, and anyone who claims to believe such things, he declares, must be either ignorant or "in the grip of faith." Little does he seem to recognize the extent to which he and his fellow scientific materialists are in the grip of their own kind of faith.

In the course of refuting some of the assumptions of scientific materialism, Searle tries to distance himself from this worldview; and he likens his enterprise to that of an anthropologist trying to describe the exotic behavior of a distant tribe. However, as a fully socialized member of the community of scientific materialists, he is in fact a full-fledged member of that not-so-distant tribe; and his work, too, embodies a good deal of its "exotic behavior." Regarding himself a cognitive scientist who has been practicing this discipline since its inception, Searle concludes: "[i]n spite of our modern arrogance about how much we know, in spite of the assurance and universality of our science, where the mind is concerned we are characteristically confused and in disagreement" (247). Within the community of philosophers of mind committed to scientific materialism, John Searle and Daniel Dennett disagree on a myriad of issues, but on this point they seem to be in complete accord. In a statement mirroring that of Searle, Dennett declares that "[w]ith consciousness . . . we are still in a terrible muddle. Consciousness stands alone today as a topic that often leaves even the most sophisticated thinkers tongue-tied and confused."[5]

Insofar as Searle adopts a genuinely empirical approach to the study of the mind, he opens up innovative and provocative ways of considering this subject in a scientific fashion. But insofar as he falls back on the principles of scientific materialism—with its disregard for the actual boundaries of scientific knowledge—his own theories appear to be confused and in disagreement with themselves and the known world of personal experience. Once again, it appears that dogmatic adherence to the metaphysical principles of scientific materialism actually works against genuine scientific research into the nature and origins of mental phenomena.

We are faced here with a dilemma not unlike that encountered by the pioneers of the Scientific Revolution. Scientific materialism assumes that mental phenomena either exist as epiphenomenal functions or emergent properties of the brain or do not exist at all. When it comes to sensory and mental qualia, including cognition and consciousness, adherents of this doctrine are determined to explain conscious events solely in terms of unconscious events. If any subjective, experiential terms are left over in one's explanation, they assume that consciousness has not really been explained at all.[6] Current neuroscientific research into the role of the brain in producing mental phenomena can be conducted with the metaphysical blinders of scientific materialism, but as long as one adheres to that doctrine, all avenues of firsthand exploration of mental phenomena themselves are prohibited. Thus, to encourage the fullest possible range of empirical, scientific methods of studying the mind, a new conceptual framework is needed that will allow for the experiential study of mental phenomena in their own terms.

$$\circ \quad \circ \quad \circ \; 8$$

SCIENTIFIC MATERIALISM
The Ideology of Modernity

Scientific Materialism and the Pursuit of Happiness

During his late twenties, William James fell victim to a sense of utterly debilitating depression that was catalyzed by his medical training at Harvard University. Specifically, this despair was brought on by the view that all our mental experiences are produced solely by brain states and there is no causal efficacy in conscious states as such.[1] In this state of acedia, he felt that "we have powers, but no motives";[2] in light of scientific materialism, all things seemed insignificant, and he was overcome by a sense of the utter insecurity of life. In his own account of this experience in *The Varieties of Religious Experience* he writes of "a horrible fear of my own existence" and says that he felt utter vulnerability to every conceivable type of suffering and fear.

> I mean that the fear was so invasive and powerful that if I had not clung to scripture-texts like "The eternal God is my refuge," etc., "Come unto me, all ye that labor and are heavy-laden," etc., "I am the resurrection and the life," etc., I think I should have grown really insane.[3]

After months of suffering from such despair, his recovery was inspired by the French philosopher Charles Renouvier, whose writings persuaded him that mental causation was indeed possible; and from that time on, he took active steps to combat his affliction by psychological means.[4] In other words, when faced with the options of adhering to the closure principle or restoring his sanity, James chose sanity. Modern scientific materialists might well ask: What was in his psychological makeup or genes that caused him to respond in this way to the closure principle? But a question that may be far more to the point is: Why has everyone else who has been indoctrinated into this reductionistic worldview not sunk into a similar, debilitating depression? Did James succumb to such despair because of a peculiarly fragile psyche? Or, because of long habituation, have our sensibilities become so inured to the implications of scientific materialism that we do not respond with authentic despair?

The principles of scientific materialism presented in the abstract might seem nothing more than innocuous metaphysical assertions, but when they are introduced into human existence with the authority of scientific knowledge, their implications are anything but innocuous. Here is the familiar picture that emerges:

The physical world is the only reality. It originates wholly from impersonal natural forces; it is devoid of any intrinsic moral order or values; and it functions without the intervention of spiritual forces of any kind, benevolent or otherwise. Life and consciousness originally arose in this universe purely by accident, from complex configurations of matter and energy. Life in general, and human life in particular, has no meaning, value, or significance other than what it attributes to itself. During the course of an individual's life, all one's desires, hopes, intentions, feelings, and so forth—in short, all one's experiences and actions—are determined solely by one's body and the impersonal forces acting upon it from the physical environment. Thus, human life is inescapably subject to suffering, for all pain and misery originate from impersonal, largely uncontrollable forces of the animate and inanimate environment and from the human body. The termination of an individual's life results in the disappearance of consciousness and the utter annihilation of the individual; and eventually this is the destiny of all life in the universe—it will simply disappear without a trace. Thus, genuine freedom from suffering and its causes occurs only at death; but of course such freedom is never experienced, for death entails the total absence of experience. In short, man is fundamentally isolated in the universe; he lives on the boundary of an alien world, which is as indifferent to his hopes as it is to his suffering or his crimes. Only by accepting this view of human existence and the universe at large can humans live authentically.[5]

Although this worldview is commonly presented as thoroughly scientific in nature, none of these assertions have been verified by empirical evidence. Rather, they are direct expressions of the dogmatic principles of scientific materialism, in accordance with which virtually all modern scientific research has been conducted. Thus, any experiences or discoveries that are found to be incompatible with that ideology are presumed to be *unscientific*. This is the scientific materialists' new word for *heretical*, and they often go even further in claiming that experience and ideas that contradict their doctrine are a priori false and illogical. Thus, students and the general public are informed, *with the full authority of science*, that if they refuse to accept the validity of the worldview just described, they are either ignorant or irrational. Such has been the strategy of ideologues throughout history.

Scientists have not proved the hypothesis that no truths lie beyond the domain of science, nor have they confirmed the hypothesis that no methodologies other than those of science can expand the horizons of human knowledge. But, with a leap of faith, scientific materialism accepts both those hypotheses as if they were established facts. William Clifford, one of the more prominent nineteenth-century scientific materialists, attacked religious faith on the grounds that "it is wrong always, everywhere, and for everyone, to believe anything upon insufficient evidence."[6] If so, all scientific materialists should immediately renounce their allegiance to their dogma.

Scientific materialism essentially reduces human existence to our physical existence, and science and technology are presented as the chief (or sole) resources for providing us with physical comfort and mental well-being. When people nowadays respond to their indoctrination into scientific materialism with despair, as did William James, they can treat this affliction with an ever-growing arsenal of drugs that affect the neurophysiological basis of depression. These drugs do not cure the depression, but they do suppress its symptoms. In addition, people suffering from chronic depression may be counseled not to dwell on the dismal aspects of existence, which, as the preceding description shows, are the personal implications of scientific materialism. Thus, such modern remedies for James's despair consist of chemical suppression of the symptoms and psychological denial of their underlying source.

In recent years, proponents of eliminative materialism, including Patricia and Paul Churchland, have argued that subjectively experienced mental states do not exist, for no account of such states can be given in terms of neuroscience. Moreover, they present this theory as a fresh, astonishing hypothesis that should startle modern thinkers much as the heliocentric theory unsettled the Scholastic contemporaries of Galileo. Two things are indeed astonishing about this materialistic account of our existence: (1) that its advocates so enthusiastically embrace an unconfirmed, speculative theory that utterly denies the validity, and even the very existence, of their personal, inner life; and (2) that anyone believes there is anything fundamentally new in this updated version of materialistic reductionism. If one accepts the closure principle and the principles of reductionism and physicalism and logically follows out their implications with reference to the human mind, eliminative materialism is the inevitable conclusion, without invoking any empirical, scientific evidence at all.

I have already noted how many of these principles can be traced back to the metaphysical speculations of Greek antiquity. But comparable theories were also developed in India beginning in the seventh century BCE or even earlier by the Indian thinker Cārvāka.[7] According to his "naturalistic" (lokāyata) philosophy, everything that happens in the universe is due solely to natural processes, and there is no such thing as supernatural causation. Only the physical universe exists, and everything consists of nothing more than configurations of the basic physical elements of nature. Thus, a human being is merely a physical organism, and consciousness is nothing more than an emergent property of specific configurations of the physical elements of the body. When those configurations vanish, so does consciousness. Thus, both the closure principle and the principles of reductionism and physicalism were already conceived in India before they occurred to the thinkers of Greek antiquity.

Cārvāka presented pleasure as the ideal of life and taught that it is to be gained by the accumulation of wealth and the pursuit of sensual and intellectual enjoyments. The fine arts, he suggested, such as music, dance, and poetry, are what make life pleasant and worth living. In his view there can be only a humanistic basis of ethics, for there is nothing that is objectively right or wrong. Indian followers of this materialistic view denied the existence of any type of divinity, and they acknowledged that their doctrine accommodated only the ideals of hedonism and sensualism—ideals that they did, in fact, espouse.

Eventually, Cārvāka and his doctrine feel into disrepute in India, for he provided no viable basis for ethics in the field of human relationships; and some of his followers took this creed as license for extravagant sexual indulgence and social chaos. The doctrine's fundamental practical flaw was that instead of encouraging humans to rise to higher ethical and spiritual levels of experience, it denied the very existence of spiritual realities and encouraged people to regard themselves as mere automatons for whom so-called ethical behavior and unethical behavior consist of merely mechanistic responses to physical stimuli. Its cognitive flaw was that it provided no genuine insight into the nature or origins of consciousness and its relation to the rest of reality. Thus, as the great contemplative traditions of India came into their full strength, this doctrine fell into decline, to the point that it is now remembered as a strange and misguided aberration in the rich history of Indian philosophy.

Modern Western civilization, on the other hand, has largely turned its back on its own contemplative heritage and, under the influence of scientific materialism, adopted a worldview and ideals closely akin to those of Cārvāka. A major difference, however, is that modern materialism has been accompanied by a rapid growth of science and technology. This has greatly enriched humanity's knowledge of and control over the natural world, and many believe it has enhanced our individual chances of survival, as well as our physical security and well-being. However, during the past Century of Scientific Materialism, we have also witnessed an accelerating growth in the world's population, rampant exploitation of the earth's natural resources, and the mass destruction of entire human communities by means of modern technology. Scientific materialism, like the materialism of Cārvāka, encourages everyone to pursue the ideal of ever higher physical living standards, and with the aid of modern technology, this has resulted in a disastrous deterioration of our entire natural environment. It has become obvious that the continuing pursuit of endlessly increasing material consumption is a sure route to ecological, economic, and social collapse.

The rise of science has been an extraordinary episode in the history of humanity in which people have sought to discover the nature of reality and the way to happiness by looking outward to the physical world instead of inward, as many of the traditional religions of the world encourage. In terms of the human pursuit of happiness, traditional religions characteristically encourage satisfaction with merely adequate physical well-being, while emphasizing the quest for ever-increasing spiritual well-being. Scientific materialism, on the other hand, at least implicitly encourages satisfaction with merely adequate mental well-being while promoting the ideal of ever-increasing physical prosperity. Thus, even the subjective experiences of peace and happiness are objectified as people become fixated on the external signs of security and enjoyment. It is human nature to seek greater happiness and security, but the ideals of traditional religions—and not the ideals associated with scientific materialism—may be the only ones that can be pursued in the long run without ruining our own physical environment.

In the early days of the Scientific Revolution, pure science was conceived by many natural philosophers as the pursuit of knowledge of the natural world as a means of indirectly knowing the mind of God. This quest may be seen as a pursuit of a kind of apotheosis, in which scientists thought God's thoughts and saw with God's own vision. But now scientific inquiry has become disengaged from the pursuit of knowledge of God, and its quest for knowledge has been reduced to a pursuit of a kind of dehumanization, in which the ideal knowledge is a view from nowhere, unrelated to human subjectivity and well-being.

Applied science is regarded as the pursuit of knowledge about the natural world in order to provide humanity with creature comforts by means of gaining control of the environment, providing protection from disease, and supporting the acquisition of power and wealth. Scientific materialism assumes that when the environment and the body, and specifically the brain, are brought under control, the mind is brought under control. Hence, in order to bring about a sense of comfort and well-being and freedom from suffering and fear, scientists have sought techniques to control the environment and maintain physical health. For those situations in which these measures prove inadequate, chemists have produced a stunning array of drugs to control the mind, such as those to enable people to relax, to become mentally aroused and alert, to sleep, to relieve anxiety, to overcome depression, to counteract attentional disorders, to improve the memory, and to experience euphoria, bliss, and even alleged mystical states of consciousness. But the vast majority of such drugs cure nothing, and their desired effects on the mind last only as long as one continues to ingest them—a point hardly lost on the pharmaceutical industry, which profits enormously from this fact. With the mainstream acceptance of legal drugs for coping with psychological problems, it should hardly come as a surprise that a sizable portion of the population in the industrially developed world avails itself of illegal drugs in its pursuit of happiness and even spiritual enlightenment.

The value system that is implicit in the worldview of scientific materialism is consumerism, and the way of life motivated by that value system centers on the amassing of wealth in order to be able to consume more and more. Tragically, the overconsumption by the industrialized world, where scientific materialism is most dominant, together with its massive proliferation of nuclear, chemical, and biological weapons, is endangering our very survival as a species, which scientific materialism presents as the central driving force of life itself.

Religions have sought to cultivate in their followers faith, love, compassion, and hope and have thereby provided millions of people throughout history with a sense of meaning and a deeper, more abiding sense of inner well-being than the more transient pleasures and sense of security provided by science and technology. According to many of the contemplative traditions of the world, a yet deeper sense of inner well-being arises from profoundly calming the mind and drawing it inward. Such psychological joy emerges not from objective stimuli but from the very nature of a balanced mind. Yet it is not immutable or unconditioned, for it depends on the continued maintenance of attentional stability and vividness. While these conditions are sustained, the mind remains in a state of unstable equilibrium and inner happiness. Finally, according to some contemplative traditions, it is possible to experience an enduring state of transcendent joy by penetrating to the experience of unstructured awareness, beyond all conceptual frameworks, and beyond all sense of subject/object duality. Such supreme happiness may be likened to maturation into spiritual adulthood, and it is found beyond the very dichotomy of stimulus-driven joy and sorrow. Thus, from a contemplative perspective, scientific materialism arrests human development in a state of spiritual infancy; and when a society of such spiritual infants is put in control of the awesome powers of science and technology, global catastrophe seems virtually inevitable.

The Institutionalization of Scientific Materialism

Scientific Materialism and Political Institutions

According to textbook histories of science, the Scientific Revolution separated science from religion to insure genuine freedom of scientific inquiry, and scientists are therefore loathe to associate their research with anything that smacks of religion. Echoing this sentiment, Steven Weinberg of the University of Texas, a Nobel laureate in physics, recently commented, "I don't want a constructive dialogue with religion. I think they should remain at odds with each other."[8] However, as the preceding chapters demonstrate, this separation has never been as complete as is commonly assumed. Scientific materialists commonly contrast the freedom of scientific institutions with the ideological tyranny so often enforced by religious institutions. But this impression is simplistic.

Institutions of scientific materialism suppress all forms of intuition, reasoning, and personal experience that are incompatible with its principles, much as did the Roman Catholic Church during the medieval era. Moreover, when it has been backed by military or police authority, proponents of this ideology have created their own inquisition to silence, subdue, or destroy all those who disagree with their dogma. Although this new religion originated in Europe, it has now spread throughout much of the world, often carried along with the propagation of the socioeconomic doctrine of Karl Marx, himself an ardent proponent of scientific materialism. Marxist regimes have characteristically suppressed all other religions, in many cases destroying churches, temples, and monasteries, torturing and executing monks, nuns, and other members of the clergy, and burning religious books and sacred images. The only way for many victims of such persecution to escape torture and death has been to publicly recant those of their religious beliefs that were incompatible with the creed of scientific materialism. In Marxist countries, indoctrination into scientific materialism has frequently been state policy. Even laypeople who resist have customarily been executed or imprisoned, and those who more quietly adhere to their faith are commonly denied educational and professional opportunities.

While Christian contemplative practice has declined with the rise of scientific materialism in Europe and North America, Buddhist contemplative practice has declined under a veritable holocaust during the Century of Scientific Materialism as a result of the ideological wars waged against it by Marxist regimes in Siberia, Mongolia, China, Tibet, Cambodia, Laos, North Korea, and Vietnam. Marxists have done their utmost to annihilate all other theories and practices they deemed religious, and Buddhism has been one of the many casualties of their militant crusades.

Communist party members in today's China, for instance, are prohibited from pursuing religious activities, whether Christian, Buddhist, Muslim, or otherwise. Within the Chinese communist empire, Tibet has especially suffered under cultural and religious genocide. If the Chinese were promoting socialism alone in Tibet, there would be no need for them to suppress the spiritual heritage of this culture. After all, the Dalai Lama, the spiritual and temporal leader of the Tibetan people, is a self-professed socialist; and when Tibetan Buddhist monks settled in India after fleeing from the Chinese communist invasion of their homeland, many of them established their monasteries on the model of socialist communes. The Chinese attack on Buddhism in Tibet was motivated not so much by the socioeconomic principles of Marxism as by the principles of scientific materialism, which Chairman Mao adopted as part of his own ideology.

Tibetans have maintained a relatively continuous tradition of Buddhist contemplative practice, since the eighth century. But during the Chinese Communist occupation of their land beginning in 1949 and the subsequent Cultural Revolution, all centers for contemplative training in Tibet were destroyed, and many contemplatives were executed. A recent article by Seth Faison in the *New York Times*, quoted an official report carried in Chinese government–controlled media that described a "patriotic education program," begun in April 1997, which the Chinese government is inflicting on the Tibetan people. This program is aimed at promoting atheism and ridding Tibetans of "passive religious influence"; it declares that "all the advantages of modern media, with its immediacy and breadth" are to be used to "popularize scientific know-how and medical knowledge." Faison concludes that "[d]espite the official policy of permitting religious worship, many Chinese officials see the intense devotion to Buddhism among Tibetans as primitive and ignorant, and the opposite of modern science and technology."[9]

While Tibetans have a great deal to learn from the sciences—indeed, with strong encouragement by the Dalai Lama, science education is now being developed in Tibetan Buddhist monasteries in exile as well as schools for the Tibetan laity—a recent study of the science and math taught in Chinese-run schools in Tibet concludes that these subjects are being socially and politically constructed and defined.[10] In other words, students in Tibet are receiving an uncritical indoctrination into scientific materialism, which overshadows any knowledge they may gain concerning the sciences themselves. The Tibetan people are being admonished that they must stop looking to their religion to solve their problems and place all their hopes in science and technology.[11] Thus, the mere fact that communist regimes endorse nothing that they call *religion* in no way prevents them from imposing the same type of intolerance commonly associated with religious institutions.

Scientific Materialism and Scientific Institutions

It is commonly assumed that modern democracies now enjoy unprecedented freedom of inquiry and belief, and that because of its twentieth-century alienation from its Christian heritage, science is now freer than ever of the constrictions of religious dogmas. However, the domination of scientific inquiry by scientific materialism suggests that this is far from the truth. With the modern institutionalization of scientific materialism, proponents of this doctrine commonly condemn even sound scientific inquiry that may undermine their creed, and they reject in principle any claims to knowledge or valid experience, especially religious experience, that is incompatible with their metaphysical assumptions.

The greatest scientific challenge to scientific materialism has come not from the cognitive sciences but from quantum mechanics. Despite the ground-shaking problems raised in this discipline, contemporary undergraduate and graduate education in physics in general and quantum mechanics in particular tends to gloss over these challenges to scientific materialism and to focus largely on the details of quantum theory itself and its practical applications in scientific research and technology. Moreover, while the philosophical challenges raised in quantum mechanics pertain to the entire natural world of micro- and macro-objects, these challenges have been largely ignored in other fields, for example, the life sciences and the cognitive sciences. Despite quantum theory's fundamental scientific challenges, proponents of scientific materialism have been able to confine these anomalies to the quantum realm; and most scientific research is still conducted within the parameters of their doctrine.

The ideological domination of scientific materialism is particularly evident in modern medicine. According to scientific materialism, there should be no reason to expect subjectively experienced mental processes such as trust, faith, belief, and expectation to exert any influence on the body. But practicing physicians have found no single, more powerful, or more ubiquitous element in healing all manner of diseases than the so-called placebo effect. One of the most renowned instances of this effect is recorded in the story of "Mr. Wright," who was diagnosed in 1957 as having cancer so advanced that he was given only a few days to live. After learning that scientists had discovered a horse serum, Krebiozen, that appeared to be effective against cancer, and after begging his physician for this medication, he was injected with this serum. Two days later, his physician found that his tumors, which had been the size of oranges, had simply vanished. Two months later, Mr. Wright read medical reports that the horse serum was a quack remedy, and he suffered an immediate relapse. His physician then injected him with a placebo, which he told his patient was "a new super-refined double strength" version of the drug; and for another two months, Mr. Wright remained in excellent health. Then he read a definitive report stating that Krebiozen was worthless, and he died two days later.

Studies have repeatedly shown that placebos can work like "real drugs," even producing side effects such as itching, diarrhea, and nausea. They have also been found to work 55–60 percent as effectively as most active medications like aspirin and codeine for controlling pain, and a recent study by psychiatrist Irving Kirsh at the University of Connecticut indicates they work about as well as modern drugs in alleviating clinical depression. Beliefs and expectations are somehow able to act like a guidance system that initiates radical and abrupt changes in both mental and physical processes, corresponding to the contents of those subjective mental states, in ways that remain unexplained within the ideological parameters of scientific materialism.

Even the name of this effect seems to be influenced by scientific materialism, for a placebo, by definition, is a harmless, unmedicated preparation given as a medicine to patients either to humor them or trick them into believing they are taking actual medication. Thus, by definition, *there can be no therapeutic effect from a placebo*! This definition itself camouflages the fact that it is not the *placebo* but the *mental processes* that have such a profound effect on human health. Scientific materialists, having no way to explain how epiphenomenal mental states could have such profound effects on the body, quickly counter that it is not the qualia of subjective mental processes that exert such influence but their "underlying" neurophysiological processes. Using new techniques of brain imagery, scientists are now discovering a host of biological mechanisms that enable placebo effects to occur. One physicalist way of interpreting the mind/brain relationship in the placebo effect is to declare that thoughts, such as beliefs and expectations, actually *turn into* the physical agents of change in the cells, tissues, and organs. According to this interpretation, thoughts themselves are defined as "a set of neurons firing which, through complex brain wiring, can activate emotional centers, pain pathways, memories, the autonomic nervous system, and other parts of the nervous system involved in producing physical sensations."[12] Thus, instead of providing an intelligible account of how subjectively experienced thoughts can influence the body, scientific materialists *define the problem away* by reducing them to objective physical processes, with no compelling logical or empirical justification whatsoever.

If the placebo effect could be reduced to some physical substance or mechanism, the production of that biological phenomenon would be a multibillion dollar industry. But since this is not the case, and perhaps because it is commonly associated with religious faith and belief, far more effort is exerted to exclude the placebo effect from "genuine" medical research than to discover the exact nature of the therapeutic efficacy of specific states of consciousness. And relatively little scientific research has been devoted to exploring how people might enhance the power of their own consciousness to induce the placebo effect more frequently and effectively. Thus, the taboo of subjectivity is held in place even at the cost of public health; and in the process, a safe distance is maintained between medical science and all religions other than scientific materialism.

Adhering to the principles of scientific materialism, many modern cognitive scientists deny the very possibility of introspection as a form of metacognition, or inner perception of mental phenomena, and they marginalize the value of introspection. Modern clinical psychologists, on the other hand, have found evidence that the loss of such self-monitoring is more damaging to the personality than the loss of a sensory faculty or motor functions. Specifically, self-monitoring is critical in acquiring and maintaining complex types of behavior and in adapting to changing conditions.[13] The very fact that we often know a lot about our present mental states, including the level and quality of our awareness, is evidence for the existence of metacognition, or introspection. However, science has yet to determine precisely which mental phenomena can be monitored and which cannot, the types of errors to which such perception is prone, how it is activated and deactivated, and how the faculty of metacognition varies from one person to the next. The recent psychological theory of "emotional intelligence" also proposes that self-monitoring of one's own mental states is a fundamental element in human intelligence and well-being; but thus far, virtually no scientific research has been conducted to determine whether this faculty of mental perception can be refined and deepened.[14] Once again, the doctrinal prohibitions of scientific materialism have obstructed and delayed scientific research that may be vital to our mental health and to scientific understanding of the mind.

Scientific materialism, like any other ideology, has gained its intellectual dominion by its conquest of institutions that shape society. Perhaps its greatest triumph has occurred with respect to modern, secular education. The First Amendment to the United States Constitution, "Congress shall make no law respecting an establishment of religion, or prohibiting the free exercise thereof," has been interpreted to mean that government schools in the United States are prohibited from teaching any religion and must on no account adopt any creed as the state religion. Scientific materialism has avoided this legality by conflating itself with scientific knowledge, distancing itself from all other religions, and by claiming that it is neutral to, and incommensurate with, religious beliefs.

By means of this strategy, scientific materialists have been extraordinarily successful in making their creed the de facto state religion of the United States. The extent of their success can be determined by examining the curricula of state schools from kindergarten through graduate school. Clearly, the world's religions have always exerted a powerful and often ubiquitous influence on the development of human communities. To understand any society, its internal functioning, and its relation to other societies, an understanding of its religious beliefs, practices, and institutions is indispensable. However, in American primary and secondary schools education in the world's religions tends to be marginalized. According to the present interpretation of the Constitution, public school teachers are not allowed to promote the truth of any religious doctrine. This is indeed a valuable safeguard against state-sponsored, religious indoctrination in a nation in which students come from a wide variety of religious backgrounds. However, this injunction should not prevent instructors from presenting the world's religious doctrines and practices as something to be taken seriously and treated with respect. Students might even be encouraged to learn *from* the world's religious traditions and not only *about* them. In today's classrooms, however, the world's religions are often so overlooked that graduating high school students commonly have only the vaguest notion of any of the world's religions unless they happen to have been brought up in a religious household and taught their parents' religion. But the same students will have become well indoctrinated into the metaphysical principles of scientific materialism, without ever being shown the distinction between this doctrine and genuine scientific knowledge. Thus, the so-called separation of church and state has resulted in students being educated in one religion only, while leading them to think of this creed as being fully validated by the authority of science.

Many state colleges and universities in the United States do not have departments or programs for the study of religion, and many do not even offer classes in religion; but it is hard to escape higher education in scientific materialism, even though it is virtually never taught as a topic in its own right. Some state universities do indeed have religious studies departments, but they have the odd distinction of being the only academic departments that are prohibited by law from promoting the truth of their subject matter, except insofar as they report the truth of what *other* people believe and practice. Religious studies is also the only academic field in which it is commonly assumed that those who neither believe in nor practice their subject matter are better able to understand it and teach it than those who do. Indeed, some religious studies departments refuse to hire anyone who has deeply held religious beliefs.

Because of this domination of scientific materialism, many scholars of religion do not dare to admit that they might actually believe in a religious worldview.[15] Moreover, the assertion that contemplatives in particular may be onto something real and valuable is commonly regarded in the academic community as somewhat disreputable, unrigorous, and unscientific. Finally, if such scholars reveal that they themselves have a regular spiritual practice and have had contemplative experiences, they open themselves up to academic ridicule on the grounds that they are being hopelessly subjective and uncritical. While scientists, historians, philosophers, and other academics are free to do their best to convince their students and colleagues of the validity and worth of their insights, scholars of religion are prohibited from promoting the truths of their own religious insights.

If scientific materialism can be advocated in state universities in the United States, should this ideological domination be broken by allowing advocates of other creeds to promote their views as the real truth? Is the secular university to be a place of multiple advocacies, or should it insist that all instructors simply present ideas and theories for students to evaluate, without advocating the truth of any of them? One might propose that university professors be allowed to advocate only a *methodology* of critical inquiry but not the truth of any ideology itself. But can one really separate methodologies from the beliefs and values that motivate them? In other words, is it even possible not to be an advocate (at least of certain values), explicitly or implicitly, if one is a teacher? Even if one advocates only a methodology of critical inquiry, one is in fact promoting a certain type of university education, and there must be a belief system that underlies that value judgment. The most feasible option, to my mind, is to insist that teachers at all levels of education make a strong effort to differentiate the *facts* of their subject matter from the ideologically driven *interpretations* of them. This, of course, is a difficult task both in the sciences and the humanities, but if this challenge is not taken seriously, ideological indoctrination is bound to replace the free inquiry that is central to a genuine liberal arts education.

Although some scholars of religion still hold to the view that only those who do not adhere to any religion are capable of scholarly research in the field of religion, this view is happily on the decline. In the meantime, some academic departments of religious studies are forums for remarkably open-minded, pluralistic, nondogmatic discussions of the nature of humanity and our relation with the rest of the universe. Teaching in the sciences, in contrast, normally entails an indoctrination into the principles of scientific materialism; and the many philosophical problems concerning the relation between scientific theory and reality are often ignored altogether. Even in philosophy departments, faculty positions are often filled by scholars whose views are compatible with those of the chair and other senior members of the department. Thus, in many such departments ideological conformity seems to be a higher priority than intellectual diversity. Especially in the fields of the philosophy of mind and the philosophy of science, scientific materialism is the prevailing ideology; and anyone who seriously challenges this dogma may find it extremely difficult to be admitted as a graduate student; or if one makes it through graduate school, the prospects for academic employment may be very dim.

Scientific and Religious Discourse

The established guidelines for teaching religion in state colleges and universities parallel those laid down by William Christian, a prominent philosopher of religion, for suitable conversation among proponents of different religious doctrines.[16] In the setting of interfaith dialogue, he suggests, people may define and explain the doctrines of their traditions, but they are not in a position to assert their validity. Even if proponents of a religion are reasonably sure that their beliefs are true, and even if they think there are valid and conclusive arguments for their validity, their religious assertions still cannot be taken as informative utterances. They are not in a position to tell their listeners what is the case, and their audience is not compelled to accept their assertions as being true. All such speakers can legitimately do is to propose beliefs.

These may seem to be reasonable guidelines set down to avoid pointless, dogmatic confrontations among people of different ideologies. But there are other issues involved that make these guidelines highly suspect, and these have to do with the incommensurability Christian asserts between religious and scientific theories. Religious doctrines, he says, are not scientific theories, for they do not present exact formulations of uniformities said to hold in the apparent world or explanations and predictions derived from these laws of nature. Although religious traditions might include among their subsidiary doctrines theories purporting to be scientific claims, he argues, these are radically unlike genuine scientific theories. The reason for this is that all major religions originated in prescientific eras, and when they have come into contact with modern science, they have learned, sometimes by bitter experience, to withdraw their purportedly scientific theories from their doctrinal schemes. And so they should, he counsels, for religious doctrines deal with a different range of problems from those that scientific theories deal with.

While it is certainly true that some problems, such as the structure of DNA, are uniquely scientific ones, and others, such as the nature of the

Trinity, are uniquely religious ones, science and religion are both vitally concerned with the human mind and its relation to the rest of nature. Moreover, many of the contemplative traditions of the world do in fact present exact formulations of uniformities said to hold concerning the human mind; and they also provide experientially-based explanations and even predictions derived from these uniformities. Thus, at the points at which religious theories do approximate scientific theories, we may ask: are religious advocates now allowed to present their views as being true, or is this still prohibited on the grounds that they do not conform to the principles of scientific materialism?

Do the same rules of engagement apply as much to the case of dialogue among advocates of scientific materialism as to that of adherents of traditional religions? As I have noted at many points already, scientific knowledge and the metaphysical assumptions of scientific materialism are commonly conflated, both in institutions of learning and in the popular media. Moreover, all the basic principles of scientific materialism, like those of other religions, originated in prescientific eras. Therefore, even if proponents of this creed are reasonably sure that their beliefs are true, and even if they think there are valid and conclusive arguments for their validity, their metaphysical assertions and interpretations of scientific evidence should, presumably, not be taken as informative utterances—at least when they are addressing people who hold other religious beliefs. Many students from grade school through university, as well as much of the public at large, do hold religious beliefs that are incompatible with the doctrine of scientific materialism. Thus, when teachers, professors, and journalists interpret scientific evidence in accordance with scientific materialism, should they not present their conclusions merely as proposals for belief and not as statements about reality? In fact, there are no restraints on their presenting their beliefs as indisputable, scientific facts; and as students and the general public receive this indoctrination, many feel that their religious beliefs have been undermined.

The differences in guidelines for religious and scientific discourse may be defended on the grounds that scientific knowledge, in contrast to religious belief, is based on experience. But many religious people strongly assert that they have experientially confirmed at least some elements of their doctrines; and even Christian acknowledges that informative utterances about the speaker's experiences may be made in the interreligious setting he proposes. If so, is it legitimate to make factual claims on the basis of one's contemplative experiences or those of the great contemplatives of one's religious tradition? Scientists commonly invoke the authority of their own experience and that of other scientists of the past and present; while advocates of scientific materialism go a step further by invoking the authority of scientists of the future, who, they believe, will discover conclusive evidence in confirmation of their metaphysical beliefs. Are religious people equally justified in calling on the experiences of the great contemplatives of their traditions?

According to the dictates of scientific materialism, all such claims are inadmissible in principle, for they flagrantly violate the taboo of subjectivity. This position is taken by Steven Katz, a contemporary scholar of religion. Katz denies that any factual propositions can be made on the basis of contemplative experience; and he insists that there can be no way of showing that contemplative experiences are true even if they are, in fact, true.[17] He maintains that the beliefs, symbols, and rituals of contemplatives define in advance what experiences they want to have and what they will be like when they have them. In other words, contemplative practice entails no genuine inquiry or observation, for it is nothing more than self-inflicted indoctrination.

Like so many proponents of scientific materialism, Katz claims he has no particular dogmatic position to defend and his writings are not based on any a priori assumptions about the nature of reality. However, one of his central theses is that there are no grounds for asserting that contemplatives ever have any conceptually unmediated experiences; and he defends his position by declaring:

> the kinds of beings we are require that experience be not only instantaneous and discontinuous, but that it also involve memory, apprehension, expectation, language, accumulation of prior experience, concepts, and expectations, with each experience being built on the back of all these elements and being shaped anew by each fresh experience.[18]

Contemplatives of diverse religious traditions commonly acknowledge that conventional experience, including scientific experience, is commonly structured by our memories, expectations, language, and so on. But many declare on the basis of their own experience that contemplative training can provide access to modes of experience that are free of all such conceptual mediation. It is certainly inappropriate to accept uncritically such experiential claims at face value; but it is equally inappropriate to deny, as Katz does, that all such claims even count as evidence simply because they violate one's a priori, dogmatic assumptions about the nature of the human mind. Katz asserts that such experience is impossible simply because of the sort of beings we are; and his coup de grâce is the assertion that "if words mean anything my position seems to be the only reasonable one to adopt."[19]

To contemplatives' experientially based challenges to the assumptions of scientific materialism Katz responds with sheer dogma and carefully selected evidence in support of his a priori assumptions. This confrontation between contemplatives and Steven Katz, as one more defender of the faith of scientific materialism, resembles that of the pioneers of modern science and the defenders of medieval Scholasticism. Contemplatives commonly insist that the only way to investigate their claims conclusively is to put them to the test of personal experience. But like the clergymen who refused to peer through Galileo's telescope, Katz claims that contemplatives themselves do not have a privileged position even when it comes to understanding the nature of their own experiences.[20] The only thing that counts as valid data for the study and analysis of contemplation, he says, are contemplatives' *accounts* of their experience; and these are equally accessible to contemplatives and scholars alike.

The Enforcement of the Taboo of Subjectivity

Throughout this work, I have suggested scientific materialism's taboo against subjectivity has curtailed scientific research into the nature, origins, and potentials of consciousness. As John Searle suggests, the "terror" of subjectivity displayed by modern scientists and scholars may be due largely to a fear of religion; and this may also account for the irrational, antiempirical dismissal of introspection as a means of acquiring scientific, firsthand knowledge of the mind. If the great contemplative traditions of the West and East have discovered avenues of insight into the nature of consciousness, these should be open to genuine empirical research—despite the misgivings of both religious and scientific dogmatists. For the curtailing of free inquiry is due to ideological taboos of all kinds and not to traditional religions alone.

The central rationale for denying in principle the validity of introspection and contemplative inquiry is that they are intrinsically subjective. Genuine observation, insists scientific materialism, requires that the object exist independently of the subject; but this notion of observation suppresses the ubiquitous fact that a subjective observer is part of the process of perceiving, identifying, and understanding *any* object. The implication of this metaphysical stance is that the validity of our knowledge of an entity is inversely proportional to the role played by subjective awareness in ascertaining its existence.

Scientific materialism assumes that the objects of scientific experience must be capable of being perceived by every competent observer; this assertion, in turn, is based on the assumption that it is impossible for an individual to develop exceptional or extraordinary perceptual abilities. But this possibility is the central claim of the contemplative traditions of the world—a claim that is ruled out *in principle* by scientific materialism. Likewise, in this scheme, every perception that deviates from "normal" is automatically deemed abnormal and therefore invalid.

The requirement just described for objectivity would fail even in bona fide scientific research, for the full significance of objects of all but quite primitive scientific observations is accessible only to individuals with very specialized training. Everyone else, including other scientists, is expected to accept those objects out of their *faith* in the integrity of science. Although the objects of contemplative experience are more private, the difference is one of degree and not of type—that is, not objective versus subjective, as commonly assumed. Scientists are no more capable of proving the validity of their most sophisticated theories to untrained (and even skeptical) people than contemplatives are able to prove theirs. Scientists are quite right to place their faith in their predecessors in their own specialized fields, in scientists in related fields on which their work relies, and in engineers who design and create their research instruments. But all such faith is combined with a healthy sense of pragmatism; and it is not blind. Such faith has an important role in both scientific and contemplative inquiry; indeed, progress in both fields may be impossible without it.

CONCLUSION
No Boundaries

The Scientific Study of Religious Experience

In a refreshing departure from all forms of religious and scientific dogmatism, William James proposes a science of religion that differs from philosophical theology by drawing inferences and devising imperatives based on a scrutiny of "the immediate content of religious consciousness."[1] This approach, he suggests, must be empirical rather than rationalistic, focusing on religious experience rather than religious doctrines and institutions. He elaborates on this point as follows.

> Let empiricism once become associated with religion, as hitherto, through some strange misunderstanding, it has been associated with irreligion, and I believe that a new era of religion as well as philosophy will be ready to begin . . . I fully believe that such an empiricism is a more natural ally than dialectics ever were, or can be, of the religious life.[2]

Such a science of religion, he suggests, "can offer mediation between different believers, and help to bring about consensus of opinion";[3] and he pondered whether such a science might command public adherence comparable to that presently granted to the physical sciences. "Even the personally nonreligious might accept its conclusions on trust, much as blind persons now accept the facts of optics—it might appear as foolish to refuse them."[4]

By the end of the nineteenth century, many physicists were utterly convinced that there were no more great discoveries to be made in their field: understanding of the physical universe was in all important respects complete. One of the few lingering problems to be solved, known as the ultraviolet catastrophe, had to do with the incompatibility of entropy-energy formulas derived from classical thermodynamics. The solution to this problem came from Max Planck, who thereby laid the foundation for modern quantum theory, which shook the very foundations of physicists' views of the universe.[5]

While there is certainly no comparable sense that the cognitive sciences have formulated a comprehensive theory of the brain and mind—far to the contrary!—many experts in this field have concluded beyond a shadow of a doubt that consciousness is produced solely by the brain and that it has no causal efficacy apart from the brain. The fact that modern science has failed to identify the nature or origins of consciousness in no way diminishes the certainty of those scientific materialists. When empirical knowledge of the nature and potentials of consciousness replaces these current metaphysical assumptions, I expect the "problem of consciousness" will turn out to have a role in the history of science comparable to that of the ultraviolet catastrophe.

The most effective way to acquire such knowledge is by a concerted, collaborative effort on the part of professional cognitive scientists and professional contemplatives, using their combined extraspective and introspective skills to tackle the hard problem of consciousness. This might entail, among other things, longitudinal studies of the gradual development of sustained, voluntary attention by people devoting themselves to contemplative training with the same dedication displayed by the scientists and engineers working on the Manhattan Project. The successful completion of those efforts to tap atomic and nuclear power changed the face of the modern world. The successful completion of such research into refined states of attention might do so as well; and if such an endeavor were pursued with the altruistic aims promoted by the great contemplative traditions of the world, the consequences for humanity may be more uniformly beneficial. This would be a refreshing departure from scientific research into mind control that has focused on controlling *others'* minds by means of drugs, other (often painful) physical stimuli, and indoctrination.

When considering the many years of arduous training and practice by accomplished contemplatives in the past, some researchers are bound to look for shortcuts, which purportedly reach the same ends with relatively little time, effort, shift of priorities, or personal sacrifice. Although it is always worthwhile to explore ever more efficient ways of pursuing knowledge, the type of contemplative research needed is one that has a parity with and complements neuroscientific research. There are no shortcuts to gaining an undergraduate and graduate education in the neurosciences, and the development of technologies that have advanced this field has been made only with long, hard work. It is unreasonable to expect that contemplative training should be any less demanding or time-consuming than scientific training; and if the former is left at an amateur level, there will be no contemplative science worthy of the name in our civilization.

There is bound to be considerable resistance among scientific materialists to such interdisciplinary and crosscultural research. Some might argue, for instance, that science should always seek new, unprecedented modes of research and not revert to prescientific theories and methods of inquiry. In particular, any collaborative association between science and traditional religions will certainly raise deep qualms. Such misgivings are based in part on the view that the great discoverers of the past had to battle against the current "facts" and dogmas of religion. Especially since the Scientific Revolution, the Western mind has sought to know what is *out there*, in the objective world, while ignoring discoveries of what is *in here*, in the subjective world. As illustrated in Daniel Boorstin's previously cited book, *The Discoverers: A History of Man's Search to Know His World and Himself*, we have commonly regarded the history of discovery as being principally a Western pursuit; and religion is commonly presented as a principal foe of discovery. Boorstin asks why other civilizations did not make various discoveries pertaining to the objective world, without asking what discoveries they did make that the Western civilization never made or has simply forgotten. The terms "mind," "consciousness," and "introspection" do not appear in the detailed index of his book, and his brief account of discoveries in the realm of the mind is confined to a few pages describing the life and works of Sigmund Freud.[6] Such ethnocentric biases concerning the history of discovery must be abandoned if we are ever to learn from the insights of the world's contemplative traditions.

With a return to empiricism as opposed to dogmatic religious and scientific rationalism, James's perspective on the future interface between science and religion is optimistic.

> Evidently, then, the science and the religion are both of them genuine keys for unlocking the world's treasure-house to him who can use either of them practically. Just as evidently *neither is exhaustive or conclusive of the other's simultaneous use*. On this view religion and science, each verified in its own way from hour to hour and from life to life, would be coeternal.[7]

With regard to contemplation, James acknowledged that while he could speak of mystical states only at second hand, he believed in the reality and paramount importance of such experiences. As a result of his studies of the experiential accounts of contemplatives from diverse traditions around the world, he concluded that

> our normal waking consciousness, rational consciousness as we call it, is but one special type of consciousness, whilst all about it, parted from it by the filmiest of screens, there lie potential forms of consciousness entirely different . . . No account of the universe in its totality can be final which leaves these other forms of consciousness quite disregarded.[8]

The great achievement of contemplatives, in his view, is the overcoming of all the usual barriers between the individual and the ultimate ground of being. This culminating experience, he believed is "the everlasting and triumphant mystical tradition, hardly altered by differences in clime or creed."[9] But he went on to say that this assertion should not obscure the real differences in contemplative methods, experiences, and theories developed among the many and diverse contemplative traditions that have appeared throughout human history. Despite this lack of unanimity, "the existence of mystical states absolutely overthrows the pretension of non-mystical states to be the sole and ultimate dictators of what we may believe."[10] Moreover, contemplatives, like everyone else, are prone to error in their understanding and evaluation of their own experiences. While scientific predictions concern future events, contemplative assertions concern events that have occurred for *others* in the *past* and that may occur for *oneself* in the *future*. Such claims by contemplatives should be taken as hypotheses, which might be ignored but cannot be refuted on rationalistic grounds. However, James believed that seriously grappling with the nature of contemplative insights might be indispensable to approaching the final fullness of truth.

It is very difficult to verify James's assertion of a single mystical state that stands at the apex of all contemplative traditions, but it is just as difficult to confirm that they are fundamentally different experiences, as Katz suggests. However, James is certainly on the right track when he proposes that the study of religious experience be empirical as well as textual and rational. Katz is presumptuous in claiming that scholars with no contemplative experience are just as competent to evaluate mystical states as are advanced contemplatives. This position is defensible only if all such states are completely conditioned by one's dogma. But since that is precisely what is to be determined, Katz's posture is incompatible with the empirical methods of science.

Some guidelines may be set down to facilitate the extraordinarily difficult scientific study of contemplative experiences. The researcher must first recognize the diverse range of contemplative experiences. Visionary experiences, which are suffused with the symbols and representations of individual traditions, must be distinguished from experiences that purportedly transcend all conceptual frameworks. While Katz seeks to validate his position by drawing on examples of the former, if there is any deep convergence of contemplative experience among diverse traditions, it is bound to occur in the latter type of insight. But research in this field should by no means be confined to an evaluation of these states themselves. No adequate study of contemplation can be conducted without a thorough investigation of its broader context of contemplative training as a whole. Thus, one must carefully examine and compare the specific forms of preparation for contemplation taught in different traditions, as well as the reported long-term effects of such experiences.

Religious creeds and practices throughout the world vary in many important respects. Therefore, particular attention should be paid to the extraordinarily similar experiential claims of contemplatives trained in different religions. As an analogy, within science, if two very different kinds of modes of observation produce similar empirical evidence, this is commonly taken to mean that researchers have discovered a phenomenon that is not simply an artifact of either mode of observation. For example, microscopes using interference, polarizing, phase contrast, direct transmission, fluorescence—all of which detect essentially unrelated aspects of light—can all be used to discern the same structures. It is unreasonable to regard this as a mere coincidence.

Generally speaking, scientists know they are observing phenomena in nature and not mere artifacts of their modes of observation by detecting them with diverse instruments and modes of experimentation. Phenomena that are suitable for scientific inquiry may be defined as scientifically noteworthy, discernible things, events, and processes that occur regularly under specifiable circumstances. According to scientific materialism, such phenomena must be recordable by competent observers, such as astronomers observing the stars, who do not substantially interfere with the objects under examination. But as philosopher Ian Hacking points out, most of the phenomena of modern physics, such as subatomic particles created in a linear accelerator, are manufactured. Thus, while the phenomena of physics are the keys that may unlock the universe, as Hacking comments, "[p]eople made the keys—and perhaps the locks in which they turn."[11]

Similarly, many contemplative experiences—especially those incorporating the symbols and representations of specific traditions—do appear to have been manufactured at least in part by those traditions. Like so many of the phenomena of physics, they, too, may be keys to unlock the mysteries of the universe; and the locks themselves may also be of our manufacture. Moreover, many contemplatives around the world and throughout history have also been sensitive to the possibility that the phenomena they experience may be mere artifacts of their own religious background. However, it has been reported that even in one contemplative tradition different methods may yield the same insight, occurring regularly under specific circumstances.[12] Moreover, especially in some advanced states of contemplation, one seeks not to interfere with the phenomena that present themselves in experience.

Throughout history, traditional religions have commonly emphasized their differences with other faiths, with many of them claiming that they alone provide the sole way to liberation, salvation, or enlightenment. While we have much to learn from these ancient traditions, perhaps modern scientific study of contemplative experiences can bring a fresh perspective to the comparative study of religion. It might even help lead to a synthetic understanding of how these diverse traditions relate to each other, while pointing out areas of contemplative experience where there is profound convergence. Truths that are invariant across the conceptual frameworks of diverse religions and even science itself may turn out to be the most significant truths available to humanity.

Scientific and Contemplative Uses of Language

Even if one grants the similarities between scientific and contemplative inquiry, at least one major difference remains: scientific discoveries can, by definition, be articulated; while many contemplatives say their discoveries are ineffable. Thus, if accounts of certain contemplative experiences cannot be taken literally, one may conclude with Steven Katz that it is logically impossible to establish any theory whatsoever on the basis of such accounts.[13] This conclusion, however, is misleading, for it ignores the diversity of ways in which language is used.

In contemplative discourse, language is often more evocative than descriptive. The distinction between descriptive and evocative language should not be regarded as absolute. The statement "Macintosh apples are red" is descriptive of Macintosh apples, but it also evokes for those familiar with apples and the color red the mental image of apples that are red. Likewise, the statement "Close the door" may evoke the intention to close the door, but that message could not be conveyed without the listener being familiar with descriptions of doors and the act of closing them. In this regard, purely descriptive and purely evocative statements do not exist as separate, independent categories. Rather, this classification describes a spectrum of speech ranging from the more explicitly descriptive to the more explicitly evocative.

Much evocative use of language by contemplatives is designed to help listeners disengage from their accustomed, conceptually structured modes of experience. While language is used evocatively by everyone, this contemplative use of language is unique in that it is intended to be self-annihilating, for it is aimed at undermining and severing the employment of all language, including itself.[14] Rather than evoking combinations of memories of previous knowledge, it is aimed at breaking through all conceptually structured experience to a radically unprecedented mode of unmediated awareness. Such discourse commonly employs metaphors and terms of negation, and their usefulness is pragmatically judged in terms of their effectiveness in leading listeners beyond the realm of conceptually mediated experience. To use a Buddhist metaphor, such verbalization is like a finger pointing to the moon: the attention must be directed beyond the finger if this pointer is to serve its purpose.

Although the highest stages of conceptually unstructured, contemplative experience are commonly said to be ineffable, there are many records of people acquiring such experience immediately upon reading contemplative literature or listening to instructions. In the Tibetan Buddhist contemplative tradition, such people are called simultaneous individuals, for they gain realization immediately upon hearing contemplative instructions.[15] Their immediate access to the ineffable is said to be due to their extraordinary degree of spiritual maturity. The same set of written or oral instructions may evoke an experience of conceptually unmediated awareness in one person and not in others, depending on their contextual knowledge and receptivity. Thus, in some sense the "ineffability" of contemplative experience seems to be relative to the listeners. By the same token, the redness of Macintosh apples is ineffable to those who are born blind.

Although scientific and contemplative discourses are certainly distinct in many important ways, the differences may easily be exaggerated. Even among scientists, learning is acquired not only by way of a set of formal instructions but also by the transmission of contextual, at times unconscious, knowledge that has come to be called tacit knowledge.[16] For instance, when the first lasers were built, written instructions on their construction proved insufficient to enable others to manufacture their own replicas.[17] The transference of tacit knowledge was found to be necessary; that is, skill in creating lasers had to be passed on from one accomplished practitioner to another. One may argue, of course, that those particular instructions were inadequate and that better written instructions would have sufficed. However, at the stage of technology in the mid–twentieth century, there may not have been enough contextual knowledge among engineers for *any written instructions* to give them sufficient information to build lasers on their own.

Like all skills, scientific and contemplative skills seem most easily acquired and developed with practice under the guidance of more experienced practitioners. When constructing a scientific apparatus or cultivating a contemplative ability, the only criterion for success is that the apparatus—be it physical or mental—finally functions as intended. Scientific and contemplative writings may give the impression that success in their respective fields comes simply by following algorithmlike instructions; and thus, carrying out their experiments may appear to be simply a formality. But as soon as difficulties arise in their work, all such pat notions are thrown to the wind, and the aid of more experienced practitioners is sought.

Both scientists and contemplatives seek to ascertain experientially whether certain hypotheses hold true. For example, physicists cannot know whether gravity waves exist until a good detector is built and it provides the correct outcome. As H. M. Collins remarks in his provocative discussion of this topic, "[i]f there are gravity waves a good apparatus is one which detects them. If there are no gravity waves the good experiments are those which have not detected them."[18] To break this circle, one must find criteria that are independent of the output of the experiment itself. Likewise, contemplatives cannot know whether conceptually unstructured awareness exists until they engage in an effective discipline that provides the correct outcome. But if such awareness is impossible, effective forms of training should demonstrate that. To escape from this circle, other criteria, such as the lasting trait-effects of contemplative practice, are invoked. Once again, it is insufficient to judge the validity and value of contemplation simply in terms of its transient state-effects. Rather, one must look as well to its overall transformation in the life of individuals and their influence on society as a whole.

One primary difference between scientific and contemplative inquiry may still seem to separate them in a most fundamental way: scientific discoveries are objective and public, whereas contemplative discoveries (if any exist) are subjective and private. While there is certainly a basis for drawing this distinction, it too may be exaggerated. When encountering a scientist's findings—even if we are scientists—most of us don't *know* that her empirical data are sound, rather we tend to take them on faith. Otherwise, the only way to know they are sound is to create a comparable laboratory of our own (we can't use hers, for if the data can be replicated only in hers and in no other laboratory, they are suspect), replicate the experiment, and see whether our findings corroborate hers. Likewise, we don't know that her mathematical analysis of her data is sound unless we apply our own analysis and thereby confirm her results. Likewise, we don't know that her theoretical interpretation of the quantitative results is sound unless we apply our own knowledge of the theory to corroborate hers. In other words, her findings—which on the surface seem to be public and third-person—are *known* by us to be valid if and only if we pursue the same research ourselves. That is, all "third-person" or collaborative research really consists of multiple first persons doing their own research and trusting the work of their collaborators.

If scientists were so skeptical of each others' work that they felt they needed to replicate any findings on which they were going to rely, scientific progress would slow to a snail's pace. But all the points in the preceding paragraph are equally true of contemplative research. Within a society in which contemplative inquiry is deemed valid and useful, most people simply trust in the authenticity of their society's greatest contemplative minds of the past and present. Professional contemplatives, while trusting the findings of their greatest predecessors and fellow contemplatives, know they must pursue the same research themselves in order to bring about the desired insights and pragmatic benefits in their own lives. These pragmatic benefits are complementary to those of technological science, as are the modes of inquiry themselves.

Faith, Contemplation, and Society

Whatever parallels may be found between scientific and contemplative inquiry, science itself is not a religion; for, unlike a religion, it provides its followers with no timeless truths on which they can rely with absolute confidence. But it appears to be an element of human nature that many followers of science, like other people, need such a firm bedrock of belief, and many of them find it in the principles of scientific materialism. In some fields of science, such as classical physics, this doctrine provides little or no impediment to the furtherance of scientific knowledge. But in the cognitive sciences it does. Specifically, the domination of scientific materialism has prevented the development of a rigorous discipline of first-person investigation of mental phenomena. This is an indispensable complement to behavioral and neuroscientific studies of the mind, regardless of one's philosophical views of the mind/body relationship. More generally, scientific materialism has stifled inquiry into the full range of mental phenomena, for it refuses to take seriously any empirical data that are incompatible with its own metaphysical assumptions.

While science, pursued within the parameters of scientific materialism, has in some respects aided us in our struggle for existence, it provides human existence with no ultimate meaning. Although many people in the modern world try to imbue their lives with religious values without questioning their often unconscious commitment to scientific materialism, such attempts are undermined from the outset. For one's values are groundless unless they are derived from one's beliefs about the very nature of reality and human existence. Traditional religions have provided generations of humanity with a sense of meaning, but the weakness of scientific materialism stems from the fact that it has no such spiritual power. At the same time, religious doctrines that fly in the face of genuine scientific knowledge are also undermined. Thus, a pressing question for our modern world is: does a way exist to integrate the power of religion and of science for the physical, mental, and spiritual well-being of humanity?

If such an integration is to occur, it is imperative that we take heed of the distinctions among science, scientific realism, scientific materialism, and scientism; for when these are conflated, fruitful association with any traditional religious outlook is precluded. Likewise, it is just as crucial to make a comparable fourfold distinction with respect to religious practices, religious philosophies, religious doctrines, and religious fundamentalism, as follows.

1. Religious practices cover a wide range of activities, such as ethical behavior, prayer, worship, and contemplation. Like the practice of science, all such practices are theory laden; yet similar activities may be pursued in conjunction with different religious philosophies and doctrines.

2. Religious people throughout history have adopted a wide array of philosophical views, which they have integrated with their practices and general doctrines. Augustine, for example, integrated much of Plato's philosophy into his understanding of Christian practice and theory, while Aquinas gave a Christian interpretation to many of the philosophical views of Aristotle. Within Indian Buddhism as well, Buddhaghosa adopted a form of substance dualism comparable to that of Descartes; Asaṅga professed a form of idealism comparable to that of George Berkeley; Padmasambhava advocated a type of conceptual relativism having points in common with Hilary Putnam's pragmatic realism and James's radical empiricism. All these philosophies are based in part on contemplative experience, but, like scientific realism, they also influence what type of knowledge is to be sought. Moreover, all

these Buddhist philosophical schools share in common a wide range of contemplative practices, such as techniques for enhancing attentional stability and vividness.

3. Religious doctrines provide stable world views and values for religious communities and fields of study for theologians and other scholars. Moreover, such theorists become the recipients and analysts of the knowledge of contemplatives and other religious adepts. Living religious traditions begin to degenerate when their followers replace effective spiritual purification, attentional training, and contemplative inquiry with sterile liturgies, ritualistic meditations, and contemplative exercises pursued with the sense that the practitioner already knows their outcome. When religious believers forsake practice altogether and simply write about others' experiences, spirituality passes from the realm of the living; and it disappears altogether when even religious people deny the validity of their own spiritual heritage.

4. The final gasp of a religion that has forsaken its contemplative heritage is fundamentalism, which throws logic and experience to the winds and defends its beliefs with a raw appeal to authority. All forms of fundamentalism, religious and scientific, regard themselves as self-sufficient, displaying no interest or concern for external challenges to their dogmas. The contamination of science with scientism and of religion with fundamentalism constitutes a lethal infection, which, if left unchecked, is bound to result in the death of its host; and the aftermath of that fatality bears little resemblance to any genuine science or religion.

Scientists and religious people alike, without exception, place their faith in some belief system that transcends the scope of their present knowledge. As William James points out, whether in scientific research or in daily life,

> we often *cannot* wait but must act, somehow; so we act on the most *probable* hypothesis, trusting that the event may prove us wise. Moreover, not to act on one belief is often equivalent to acting as if the opposite belief were true, so inaction would not always be as "passive" as the intellectualists assume.[19]

Faith, he asserts, is essential, but as a practical, not a dogmatic, attitude, and it must go with toleration of other faiths, with the search for the most probable, and with the full consciousness of responsibilities and risks. Specifically, he defends one's right to believe ahead of the evidence only in those cases where (1) much is at stake, (2) the evidence at hand does not settle the case, and (3) one cannot wait for more evidence, either because no amount of evidence can settle the case or because waiting itself is to decide not to believe. While the role of faith in both science and religion can hardly be denied, one salient difference remains: scientific theories must be capable in principle of empirical refutation; but some religious beliefs may be simply incapable of such repudiation.

Regarding the relation of contemplation to society at large, Augustine asserts that very few individuals may gain certain knowledge of the changeless Truth; so for the general laity, faith must substitute for contemplation. As long as a religion's contemplative tradition is alive, the religious community at large can look to the experiences of its contemplatives to support their beliefs, much as the modern West nowadays places its faith in scientific research. For this reason, the central importance of contemplation for society as a whole is emphasized not only in the religions of Asia, but in prescientific Christianity. Thomas Aquinas, for instance, declares that "[i]t is requisite for the good of the human community that there should be persons who devote themselves to the life of contemplation."[20] And in terms of the ordering of society as a whole, he asserts:

> the whole of political life seems to be ordered with a view to attaining the happiness of contemplation. For peace, which is established and preserved by virtue of political activity, places man in a position to devote himself to contemplation of the truth.[21]

At least since the Scientific Revolution and the Protestant Reformation, there have been relatively few individuals in Western civilization who have devoted themselves to the life of contemplation; and our culture has been the poorer for this absence. Religious writings in general, and contemplative writings in particular, may be likened to paper currency, while religious experience and especially contemplative experience are like gold reserves. To the extent that religious experience is no longer current or considered to have value, religious texts appear to the outsider to have no truth value; and an entire dimension of human existence—from a contemplative point of view, the most important dimension—is sacrificed as a result.

From a contemplative perspective, the current scientific view of the world is fundamentally flawed, for it has failed to take into account the role and significance of consciousness in nature. The reason for that is that science has not developed effective methods for exploring consciousness firsthand; and the reason for that is that scientific inquiry has been constrained by the metaphysical principles of scientific materialism. This dogma allows science to explore only those facets of reality that conform to its creed; and the experienced mind is simply left out. From a scientific perspective, religious views of the world are fundamentally flawed, for they are not evidently based on a precise, critical exploration of the natural world. The reason for that is that the world's religions have for the most part turned their backs on whatever contemplative methods they may have had for exploring reality; and the reason for that is that they have been constrained by unskeptical adherence to authority and tradition.

Contemplation invites empirical inquiry from both scientific and religious perspectives, and if it is revitalized in modern civilization—in the West and the East—it may take on a mediating role between science and religion and even among diverse religions. In the history of science many people have found scientific research to be a profoundly religious pursuit entailing self-transformation, and many others have been drawn to science simply by their zeal for understanding. Likewise, in the modern world both motivations may inspire researchers in the field of contemplative inquiry. With the restoration of subjectivity into the natural world, science and religion may challenge each other to open up new domains of experience. I believe this is what William James envisioned when he wrote:

> [t]he whole drift of my education goes to persuade me that the world of our present consciousness is only one out of many worlds of consciousness that exist, and that those other worlds must contain experiences which have a meaning for our life also; and that although in the main their experiences and those of this world keep discrete, yet the two become continuous at certain points, and higher energies filter in.[22]

Science and religion have both proven they are here to stay, at least for the foreseeable future. They may coexist in mutual ignorance of each other's insights and power; each one may try to suppress or eliminate the other; or they may finally learn that their worlds inevitably intersect, and that such areas of common ground need not be seen as a threat but may be seen as an opportunity for greater understanding. The point at which science and religion must overlap is the human mind itself; yet the origins, nature, and final destiny of the mind remain hidden from public knowledge. The empirical study of the mind, unconstrained by the dogmatic principles of scientific materialism and all other religious creeds, awaits us. We are faced with the challenge of restoring our own subjectivity to the natural world, acknowledging its meaningful role in nature. The methods of both science and religion provide us with indispensable tools for such research; and, as William James suggests, we may find that at this point of intersection between the worlds of science and of religion, higher energies filter in.

NOTES

INTRODUCTION

1. See Güven Güzeldere (1995).
2. Augustine (391/1937), bk. 3, chs. 20–21.
3. N. S. Sutherland (1989).
4. R. Shorto (1997) and Sharon Begley (1998).
5. *International Herald Tribune*, December 24–5, 1991, p. 4.
6. Bertrand Russell (1961) pp. 171–72.
7. Peter Smith and O. R. Jones (1988) p. 29.
8. See Albert Einstein (1950), p. 25, (1954), p. 40.
9. See Bronislaw Malinowski (1925/1948), and Michael A. Arbib and Mary B. Hesse (1986), p. 17.
10. See Émile Boutroux (1911) p. 324.
11. For other arguments promoting the separation of religion and science since the European Enlightenment, see Immanuel Kant (1781/1998), (1788/1997) and Friedrich Schleiermacher (1799/1988); among contemporary thinkers, see Holmes Rolston III (1987) and Stanley J. Tambiah (1990).
12. Jeremy W. Hayward and Francisco J. Varela (1992); Dalai Lama (1992); Francisco Varela (1997); Daniel Goleman (1997); Zara Houshmand, Robert Livingston, and B. Alan Wallace, (1999).
13. See Dalai Lama (1996); Rodger Kamenetz (1994).

1. FOUR DIMENSIONS OF THE SCIENTIFIC TRADITION

1. See B. Alan Wallace (1996), ch. 2.
2. See Paul Feyerabend (1994a).
3. See Bas C. van Fraassen (1989), p. 9.

4. Charles Taylor (1989), pp. 33–34.
5. Ernan McMullin (1994), pp. 103–4.
6. Arthur Koestler (1959), p. 447.
7. Bas C. van Fraassen (1989). See also Michael A. Arbib and Mary B. Hesse (1986).
8. Ian Hacking (1983).
9. See Paul M. Churchland (1990a).
10. See Paul Feyerabend (1994b).
11. See Jacques Monod (1971). For a revealing account of the many nonobjective elements that influence scientific research and knowledge, see Thomas S. Kuhn (1970).
12. Cited in D. Wilson (1983), p. 391.
13. See Richard Feynman, R. B. Leighton, and M. Sands (1963), pp. 1–9.
14. See Evan J. Squires (1990), p. 15.
15. See Peter Coveney and Roger Highfield (1990), pp. 295–97.
16. See Paul M. Churchland (1990a), p. 11.
17. Edward O. Wilson (1998a), p. 54; see Edward O. Wilson (1998b).
18. See B. Alan Wallace (1996), ch. 3.
19. See Albert Einstein (1954), "On Scientific Truth"; (1950).
20. See Thomas Nagel (1986).
21. Patricia Smith Churchland (1998), p. 127.
22. Güven Güzeldere (1998), p. 25.
23. William James (1890/1950), pp. 290–91.
24. Ibid., p. 322. Throughout this book, all italics in quoted citations are found in the original works unless otherwise indicated.
25. Cited in Arthur Koestler (1967), p. 5.
26. John B. Watson (1913).
27. Cited in Arthur Koestler (1967), p. 5.
28. Ibid., p. 7.
29. See A. J. Ayer (1946), pp. 90–94.
30. B. F. Skinner (1974), p. 4.
31. See Paul M. Churchland (1990a) and Stephen Stich (1983).
32. See Dudley Shapere (1982).
33. Edward O. Wilson (1998a), p. 65.
34. Ibid., p. 68.
35. Ibid., p. 68.
36. Ibid., p. 64.
37. See Stephen Hawking's comments in Renée Weber (1986), p. 208; Richard Feynman (1983), pp. 172–73.
38. Emile Durkheim (1912/1965), p. 56.
39. Ibid., p. 341.
40. Ibid., p. 338.
41. Ibid., p. 340.
42. Ibid., p. 224.
43. Ibid., p. 225. Retranslated by Steven Lukes (1973), pp. 464–65.
44. Galileo, "Letter to the Grand Duchess Christina," in Stillman Drake (1957), pp. 182–83.
45. Albert Einstein (1954), pp. 274, 232.

46. Ibid., p. 223.
47. Ibid., p. 270.
48. Richard Feynman, R. B. Leighton, and M. Sands (1963), p. 4–2.
49. See Sir Arthur Eddington (1955), p. 217.
50. See B. Alan Wallace (1996), ch. 3.
51. For a critique of the apparent unity of science, see Peter Galison and David J. Stump (1996).
52. Emile Durkheim (1912/1965), p. 477.
53. Cited in John Hedley Brooke (1991), p. 31.
54. Albert Einstein (1954), p. 40.
55. Albert Einstein, "Autobiographical Notes," in Paul Arthur Schlipp (1969), pp. 3–5.
56. See Peter Medawar (1984), p. 60; Douglas Sloan (1983), p. 7, Michael A. Arbib and Mary B. Hesse (1986), p. 197; and Huston Smith (1982).

2. THEOLOGICAL IMPULSES IN THE SCIENTIFIC REVOLUTION

1. See Augustine (391/1937), bk. 2, chs. 10–11 and 16.
2. See Keith Thomas (1971).
3. Exodus 22:18. *New International Version.*
4. Lorraine Nicolas Remy (1595/1930) p. xii, cited in Brian Easlea (1980), p. 29.
5. See David Ray Griffin (1989), pp. 84–86.
6. W. F. Whitehead (1897/1971), bk. 1, ch. 68, p. 206, cited in Brian Easlea (1980), p. 94.
7. Robert K. Merton (1949), p. 329.
8. René Descartes (1973), pt. 4, sec. 187; (vol. 3, p. 502n).
9. Francis Bacon, *Novum Organon*, in Spedding, vol. 4, p. 47, cited in Brian Easlea (1980), 128.
10. See Daniel 12:3–4.
11. René Descartes (1960), pt. 6, sec. 62 (p. 45).
12. See Michael Foster (1934), (1936) and Francis Oakley (1961).
13. Otto von Gierke (1927), p. 173, n. 256.
14. Jean Calvin (1989), I.v.5.
15. Thomas Sprat (1667/1959), pp. 339–41, cited in Brian Easlea (1980), p. 4.
16. See H. G. Alexander (1717/1956).
17. See Stanley Tambiah (1990), p. 17.

3. AN EMPIRICAL ALTERNATIVE

1. William James (1890/1950), pp. 290–91.
2. See Henry Samuel Levinson (1981); Alfred North Whitehead (1938), pp. 3–4; Edwin G. Boring (1950), p. 743.
3. See William James, "Radical Empiricism" (1897 & 1909) in John J. McDermott (1977), pp. 134–36.
4. William James, "A World of Pure Experience" (1912), in John J. McDermott (1977), p. 195.
5. John Anderson (1990), p. 24.
6. Michael D. Lemonick (1995).

7. See Jerome Bruner (1973).
8. See Per F. Dahl (1997) and http://wheel.ucdavis.edu/btcarrol/skeptic/blond-lot.html.
9. See Ian Hacking (1983), pp. 188–207.
10. Jerome Bruner (1986), p. 46.
11. Daniel J. Boorstin (1985), p. xv.
12. See Wilfrid Sellars (1963).
13. Hilary Putnam (1990), p. 28.
14. Ibid.
15. Hilary Putnam (1988), p. 113.
16. William James, "The Notion of Consciousness" (1905), in John J. McDermott (1977), p. 194.
17. Hilary Putnam (1991), p. 407.
18. Ibid., p. 422–23.
19. See Hilary Putnam (1988), pp. 109–119.
20. Hilary Putnam (1990) p. 30.
21. William James (1907/1975), p. 97.
22. Ibid., p. 100.
23. See Robin Horton (1982); Hilary Putnam (1987); (1990); (1991); and Thomas McFarlane (1996), (1997).
24. See B. Alan Wallace (1996), chs. 1 and 2.
25. See Henry P. Stapp (1993).
26. See Evan J. Squires (1994), pp. 201–4; Eugene P. Wigner (1967).
27. Cited in John Gribbin (1984), p. 212.
28. Cited in Jeremy Bernstein (1985).
29. See David Hodgson (1994), pp. 205–16.
30. See Hilary Putnam (1975), pp. 295–97.
31. Werner Heisenberg (1962), p. 70.
32. Niels Bohr (1987), vol. 1, p. 54.

4. OBSERVING THE MIND

1. See Keith Sutherland (1994), pp. 285–86.
2. Daniel Dennett (1969), p. 40. See William Lyons (1986) and Kurt Danziger (1980), 241–62.
3. Augustine (391/1937), book 2, chs. 3–5.
4. See René Descartes (1964).
5. See Eugene Taylor (1990), p. 56.
6. Kurt Danziger (1980), p. 259; see Gerald E. Myers (1986), pp. 64–80.
7. William James (1909/1977), p. 17.
8. Émile Boutroux (1911), p. 329.
9. Ibid., p. 338.
10. See B. Alan Wallace (1996), ch. 9.
11. Immanuel Kant (1786/1970), p. 8.
12. See K. Lashley (1956).
13. Werner Heisenberg (1962), p. 58.
14. Cited in Werner Heisenberg (1971), p. 63.
15. E. von Aster (1908), p. 65, cited in Kurt Danziger (1980), p. 257.

16. William Lyons (1986), p. 104.
17. Ibid., p. 154.
18. Ibid.
19. See H. Ginsburg and S. Opper (1979), pp. 175–77; John Broughton (1978).
20. Robert Woodworth and Harold Schlosberg (1955), p. 90.
21. William James (1911), pp. 22–24.
22. See Eugene Taylor (1990), p. 56.
23. William James (1892), p. 146.
24. Ibid., p. 185.
25. William James (1899/1958), vol. 1, pp. 28–29.
26. William James (1890/1950), vol. 1, p. 195.
27. See B. Alan Wallace (1998), pp. 278–83.
28. William James (1890/1950), vol. 1, pp. 189–90.
29. René Descartes (1960), p. 150.
30. Ibid., p. 151.
31. William James (1890/1950), vol. 1, pp. 191–92, 197–98.
32. William James (1899/1958), p. 19.
33. See B. Alan Wallace (1998), pp. 269–89.

5 · EXPLORING THE MIND

1. William James (1899/1958), p. 127.
2. William James (1890/1950) vol, 1, p. 424.
3. Ibid.
4. James Deese (1990), p. 295.
5. William James (1890/1950) vol., 1, pp. 447–48.
6. William James (1899/1958), p. 126.
7. William James (1902/1985), pp. 400–402.
8. William James (1890/1950), vol. 1, pp. 420–23.
9. William James (1899/1958), p. 84.
10. See N. H. Mackworth (1950); J. F. Mackworth (1970); A. F. Sanders (1970).
11. See M. I. Posner (1978).
12. See Esther K. Sleator, and William E. Pelham, Jr. (1986); Ronald A. Cohen (1993).
13. *Physicians' Desk Reference* (1995), p. 897.
14. James S. Hans (1993), pp. 36, 40.
15. Dom Cuthbert Butler (1967), p. 29. In this discussion I am substituting the theological term "soul" for the more neutral terms "mind" and "consciousness." Even though they are certainly not equivalent, Christian references to the soul certainly do include the mind and consciousness.
16. Owen Chadwick (1958), pp. 198, 241.
17. M. O'C Walshe (1979), vol. 2, p. 14.
18. Ibid., vol. 1, p. 7.
19. James Clark and John Skinner (1958), p. 101; Robert K. C. Forman (1990b), p. 104.
20. William James (1905/1977), p. 191.

21. Augustine (416/1962), bks. 9 and 10. See Phyllis Hodson (1955), p. 3; Justin McCann (1952), pp. 140–141.
22. Augustine (416/1962), bk. 14, ch. 6, p. 421.
23. Dom Cuthbert Butler (1967), p. xxiv.
24. See Daniel C. Matt (1995); Dom Cuthbert Butler (1967); Swami Prabhavananda and Christopher Isherwood (1981) and William M. Indich (1995); Peter Harvey (1995) and Thrangu Rinpoche (1993).
25. William James (1902/1960) pp. 367–68.
26. John Burnaby (1938/1991) p. 52.
27. M. O'C Walshe (1979), vol. 1, p. 7.
28. Buddhaghosa (1979), chs. 4 and 5; Paravahera Vajirañāna (1975), ch. 13.
29. Buddhaghosa (1979), chs. 7 and 8; David W. Evans (1992), pp. 213–14.
30. Buddhaghosa (1979), ch. 12, US. 87–91.
31. Paravahera Vajirañāna (1975), pp. 151, 327–28; David J. Kalupahana (1987), pp. 112–15; Peter Harvey (1995), pp. 155–179.
32. See B. Alan Wallace (1998); (1999); Gen Lamrimpa (1995).
33. Vasubandhu (1991), vol. 2, p. 474.
34. Ibid., vol. 1, p. 190. I have altered the translation of Poussin/Pruden slightly so that the terminology conforms to this work.
35. See Karma Chagmé (1998), pp. 80–84; Gyaltrul Rinpoche (1993), pp. 134–35, 151–54.
36. Padmasambhava (1998), pp. 105–14.
37. René Descartes (1964), p. 110.
38. William James (1904/1977), p. 169.
39. William James (1905/1977), p. 193.
40. Ibid., p. 194.
41. William James (1904/1977), pp. 177–78.
42. Ibid., p. 172.
43. Ibid., p. 176; see William James (1890/1950), vol. 2, p. 294.
44. See Gerald Myers (1986), p. 64.
45. William James (1912/1976), p. 46.
46. See Padmasambhava (1998), ch. 5.
47. Padmasambhava (1998), p. 121.
48. Ibid., p. 122.
49. See Khenchen Kunzang Palden and Minyak Kunzang Sönam (1993), pp. 49–55.
50. See Padmasambhava (1998), pp. 179–193.
51. See Donald Rothberg (1990).
52. See Robert K. C. Forman (1990c); Peter Harvey (1995); William M. Indich (1995); David Loy (1988); Karma Chagmé (1998); Franklin Merrell-Wolff (1994); (1995).
53. Robert K. C. Forman (1990b), p. 106.
54. B. Alan Wallace (1998), pp. 230–48; Gen Lamrimpa (1999); Padmasambhava (1998), 114–40; 169–193; Karma Chagmé (1998), pp. 85–123.
55. Brian D. Josephson, in Michel Cazenave (1984), pp. 9–19.
56. Evan J. Squires (1990), p. 40.
57. Nick Herbert (1985), p. 248.

6. THE MIND IN SCIENTIFIC MATERIALISM

1. Howard Gardner (1985), p. 39.
2. Ibid., p. 6.
3. See T. Shallice (1972).
4. See Francisco Varela, Evan Thompson, and Eleanor Rosch (1991), pp. 157–71; Evan Thompson (1995).
5. See Paul M. Churchland (1990a), pp. 16–18.
6. Francis Crick and Christof Koch (1998).
7. William James (1989), p. 85–86.
8. Ibid., p. 87.
9. Augustine (391/1937), bk. 3, chs. 20–21.
10. Ibid., p. 379.
11. Güven Güzeldere (1998), p. 45.
12. Bernard d'Espagnat (1981), p. 84; B. Alan Wallace (1996), ch. 6. Edward R. Harrison (1981), p. 148. Heisenberg, cited in Nick Herbert (1985), p. 22. Stapp cited in Paul Davies (1985), p. 49.
13. Richard Feynman, R. B. Leighton, and M. Sands (1963), p. 4–2.
14. See D. M. Armstrong (1990), p. 39.
15. David Hume (1980), p. 32.
16. William James (1890/1950), vol. 2, p. 291.
17. See Stephen Stich (1983), Kathleen Wilkes (1988), and Paul M. Churchland (1990b).
18. Paul M. Churchland (1990a), pp. 41, 48.
19. Daniel C. Dennett (1991), p. 74.
20. See Patricia S. Churchland (1986).
21. Paul M. Churchland (1990a), p. 46.
22. See Robin Horton (1982).
23. See William Lyons (1986), pp. 124, 149, 155.
24. See Frank Jackson (1990), p. 475.
25. See Terrence Horgan and James Woodward (1990), pp. 414, 418, n. 18.
26. Quoted in K. C. Cole (1999a).
27. John Horgan (1996), p. 8.

7. CONFUSING SCIENTIFIC MATERIALISM WITH SCIENCE

1. See Raymond Tallis (1994).
2. John Anderson (1990), p. 9. Page numbers cited in text hereafter.
3. John R. Searle (1994), p. 95. Page numbers cited in text hereafter.
4. Ibid., p. 152.
5. Daniel Dennett (1991), pp. 21–22.
6. See ibid., pp. 454–55.

8. SCIENTIFIC MATERIALISM: THE IDEOLOGY OF MODERNITY

1. Margaret Knight (1950), p. 30.
2. William James (1897/1979), p. 71.
3. William James (1902/1985), p. 161.

4. See John J. McDermott (1977), pp. 6–8.
5. See Jacques Monod (1971), p. 167.
6. William Kingdon Clifford (1879), vol. 2, p. 183, cited in William James (1897/1979), p. 18.
7. See P. T. Raju (1985), ch. 3.
8. Quoted in K. C. Cole (1999b).
9. Seth Faison (1999).
10. Liu Kiufeng (1996).
11. *Tibet Daily*, November 26, 1996.
12. Sandra Blakeslee (1999).
13. See David Galin (1992); J. H. Flavell and H. M. Wellman (1977); E. M. Markman (1977).
14. See Daniel Goleman (1995).
15. See William G. Bernard (1994).
16. See William A. Christian (1972).
17. See Steven T. Katz (1978).
18. Ibid., p. 59.
19. Ibid., p. 40.
20. Steven T. Katz (1983), p. 5.

CONCLUSION: NO BOUNDARIES

1. William James (1902/1985), p. 12.
2. William James (1909/1977), p. 142.
3. William James (1902/1985), p. 456.
4. Ibid.
5. See E. T. Whittaker (1954), ch. 3.
6. Daniel J. Boorstin (1985), p. xv.
7. William James (1902/1985) pp. 122–3. My italics.
8. Ibid., p. 388.
9. Ibid., p. 419.
10. Ibid., p. 427.
11. Ian Hacking (1983), p. 228.
12. See B. Alan Wallace (1998), p. 248.
13. Steven T. Katz (1978), p. 40.
14. See Robert K. C. Forman (1994).
15. Karma Chagmé (1998), ch. 7.
16. See Michael Polanyi (1966).
17. See. H. M. Collins (1985), pp. 51–111; Steven Shapin and Simon Schaffer (1985).
18. H. M. Collins (1985), p. 89.
19. William James, "Faith and the Right to Believe" (1911), in John J. McDermott (1977), p. 736.
20. Thomas Aquinas, *Sentences of Peter Lombard* 4d. 26, 1, 2, cited in Josef Pieper (1966), p. 96.
21. Thomas Aquinas, *Commentary on Aristotle's Nicomachean Ethics* 10, 11; no. 2101, cited in Josef Pieper (1966), p. 94.
22. William James (1902/1985), p. 519.

BIBLIOGRAPHY

Alexander, H. G., ed. (1717/1956) *The Leibniz-Clarke Correspondence*. Manchester, England: Manchester University Press.

Anderson, John. (1990) *Cognitive Psychology and Its Implications*. New York: W. H. Freeman.

Arbib, Michael, and Mary B. Hesse. (1986) *The Construction of Reality*. Cambridge: Cambridge University Press.

Armstrong, D. M. (1990) "The Causal Theory of the Mind." In *Mind and Cognition: A Reader*, edited by William G. Lycan. Cambridge, Mass.: Blackwell. Pp. 37–47.

Augustine. (391/1937) *The Free Choice of the Will*. Translated by Francis E. Tourscher. Philadelphia: Peter Reilly.

———. (416/1962) *The Trinity*. Translated by Stephen McKenna. Washington, D.C.: Catholic University of America Press.

Ayer, A. J. (1946) *Language, Truth and Logic*. 2nd ed. London: Gollancz.

Baldwin, James Mark, ed. (1925) *Dictionary of Philosophy and Psychology*. New York: MacMillan.

Begley, Sharon. (1998) "Science Finds God." *Newsweek*, July 20, 1998.

Ben-David, Joseph. (1985) "Puritanism and Modern Science: A Study in the Continuity and Coherence of Sociological Research." In Erik Cohen, *Comparative Social Dynamics*, edited by Moshe Lissak, and Uri Almagor. Boulder, Colo.: Westview Press. Pp. 207–23.

Bernard, G. William. (1994) "Transformations and Transformers: Spirituality and the Academic Study of Mysticism." *Journal of Consciousness Studies: Controversies in Science and the humanities*, 1, no. 2, pp. 256–60.

Bernstein, Jeremy. (1985) "Physicist John Wheeler: Retarded Learner." *Princeton Alumni Weekly*, October 9, 1985, pp. 28–41.

Bjork, Daniel W. (1988) *William James: The Center of His Vision*. New York: Columbia University Press.

Blakeslee, Sandra. (1999) "Placebos Prove So Powerful Even Experts Are Surprised." *New York Times*, October 13, 1998.

Block, N., O. Flanagan, and G. Güzeldere, eds. (1998) *The Nature of Consciousnes: Philosophical Debates*. Cambridge, Mass.: MIT Press.

Bohr, Niels. (1987) *The Philosophical Writings of Niels Bohr*. Woodbridge, Conn.: Ox Bow.

Boorstin, Daniel J. (1985) *The Discoverers: A History of Man's Search to Know His World and Himself*. New York: Vintage Books.

Boring, Edwin G. (1950). *A History of Experimental Psychology*. 2nd ed. New York: Appleton-Century-Crofts.

Boutroux, Émile. (1911) *Science and Religion in Contemporary Philosophy*. Translated by Jonathan Nield. New York: Macmillan.

Brooke, John Hedley. (1991) *Science and Religion: Some Historical Perspectives*. Cambridge: Cambridge University Press.

Broughton, John. (1978) "Development and concepts of Self, Mind, Reality, and Knowledge." In *Social Cognition*, edited by W. Damon. San Francisco: Jossey-Bass.

Bruner, Jerome. (1973) *Beyond the Information Given: Studies in the Psychology of Knowing*. New York: Norton.

———. (1986) *Actual Minds, Possible Worlds*. Cambridge, Mass.: Harvard University Press.

Buddhaghosa. (1979) *The Path of Purification*. Translated by Bhikkhu Ñāṇamoli. Kandy, Sri Lanka: Buddhist Publication Society.

Burnaby, John. (1938/1991) *Amor Dei: A Study of the Religion of St. Augustine*. Norwich, England: Canterbury Press.

Butler, Dom Cuthbert. (1967) *Western Mysticism: The Teaching of Augustine, Gregory and Bernard on Contemplation and the Contemplative Life*. 3rd ed., with "Afterthoughts" by David Knowles. London: Constable.

Calvin, Jean. (1989) *Calvin's Institutes: A New Compendium*. Edited by Hugh T. Kerr.

Louisville, Ky.: Westminster/John Knox Press.

Cazenave, Michel, ed. (1984) *Science and Consciousness: Two Views of the Universe*. Translate by A. Hall and E. Callander. Oxford: Pergamon Press.

Chadwick, Owen, trans. and ed. (1958) *The Conferences of Cassian in Western Asceticism*. Philadelphia: Westminster Press.

Christian, William A. (1972) *Oppositions of Religious Doctrines: A Study in the Logic of Dialogue among Religions*. London: Macmillan Press.

Churchland, Patricia Smith. (1986) *Neurophilosophy*. Cambridge, Mass.: MIT Press.

———. (1998) "Can Neurobiology Teach Us Anything about Consciousness?" In *The Nature of Consciousness*, edited by N. Block, O. Flanagan, and G. Güzeldere. Cambridge, Mass.: MIT Press. pp. 127–140.

Churchland, Paul M. (1990a) *Matter and Consciousness: A Contemporary Introduction to the Philosophy of Mind*. Rev. ed. Cambridge, Mass.: MIT Press.

———. (1990b) "Eliminative materialism and Propositional Attitudes." In *Mind and Cognition*, edited by William G. Lycan. Oxford: Blackwell. Pp. 206–23.

Clark, James, and John Skinner, eds. (1958) *Meister Eckhart: Selected Treatises and Sermons*. London: Faber and Faber.

Clifford, William Kingdon. (1879) *Lectures and Essays*. Edited by Leslie Stephen and Frederick Pollack. 2 vols. London: Macmillan.

Cohen, Ronald A. (1993) *The Neuropsychology of Attention*. New York: Plenum Press.

Cole, K. C. (1999a) "In Patterns, Not Particles, Physicists Trust." *Los Angeles Times*, March 4, 1999.

———. (1999b) "Religion vs. Science: Devil Is in the Details." *Los Angeles Times*, April 22, 1999.

Collins, H. M. (1985) *Changing Order: Replication and Induction in Scientific Practice*. London: Sage. Pp. 51–111.

Coveney, Peter, and Roger Highfield. (1990) *The Arrow of Time: A Voyage through Science to Solve Time's Greatest Mystery*. New York: Fawcett Columbine.

Crick, Francis, and Christof Koch. (1998) "Towards a Neurobiological Theory of Consciousness." In *The Nature of Consciousness: Philosophical Debates*, edited by N. Block, O. Flanagan, and G. Güzeldere. Cambridge, Mass.: MIT Press. Pp. 277–92.

Dahl, Per F. (1997) *Flash of the Cathode Rays: A History of J. J. Thomson's Electron*. Philadelphia: Institute of Physics.

Dalai Lama. (1992) *Worlds in Harmony: Dialogues on Compassionate Action*. Berkeley, Calif.: Parallax Press.

———. (1996) *The Good Heart: A Buddhist Perspective on the Teachings of Jesus*. Boston: Wisdom.

Danziger, Kurt. (1980) "The History of Introspection Reconsidered." *Journal of the History of the Behavioral Sciences*, 16, pp. 241–62.

Davies, Paul. (1985) *Superforce*. New York: Simon and Schuster.

Deese, James. (1990) "James on the Will." In *Reflections on* The Principles of Psychology: *William James after a Century*, edited by Michael G. Johnson and Tracy B. Henley. Hillsdale, N.J.: Erlbaum.

Dennett, Daniel. (1969) *Content and Consciousness*. New York: Routledge and Kegan Paul.

———. (1991) *Consciousness Explained*. Boston: Little, Brown.

Descartes, René. (1960) *Discourse on the Method*. Translated by Laurence J. Lafleur. New York: Bobbs-Merrill.

———. (1964) *Discourse on Method and Other Writings*. Translated by Arthur Wollaston. Baltimore: Penguin.

———. (1973) *Principles of Philosophy*. Translated by F. Alquié. In *Oevres Philosophiques de Descartes*, edited by Alquié. Garnier.

d'Espagnat, Bernard. (1981) *In Search of Reality*. New York: Springer-Verlag.

Dieks, Dennis. (1994) "The Scientific View of the World." Introduction *Physics and Our View of the World*, edited by Jan Hilgevoord. Cambridge: Cambridge University Press. Pp. 61–78.

Drake, Stillman. (1957) *Discoveries and Opinions of Galileo.* New York: Anchor Books.

Drees, Willem B. (1994) "Problems in Debates about Physics and Religion." In *Physics and Our View of the World* edited by Jan Hilgevoord. Cambridge: Cambridge University Press. Pp. 188–225.

———. (1996) *Religion, Science and Naturalism.* Cambridge: Cambridge University Press.

Durkheim, Emile. (1912/1965) *The Elementary Forms of the Religious Life.* Translated by Joseph W. Swain. New York: Macmillan.

Easlea, Brian. (1980) *Witch-hunting, Magic and the New Philosophy: An Introduction to Debates of the Scientific Revolution 1450–1750.* Brighton, N.J.: Humanities Press.

Eckhart, Meister. (1979 and 1987) *Meister Eckhart: Sermons and Treatises.* Vols. 1–3. Translated by M. O'C Walshe. Dorset, England, Longmead: Element Books.

Eddington, Sir Arthur. (1955) "The Domain of Physical Science." In *Science, Religion, and Reality,* edited by Joseph Needham. New York: Braziller. Pp. 193–222.

Einstein, Albert. (1950) *Out of My Later Years.* New York: Philosophical Library.

———. (1954) *Ideas and Opinions.* Translated by Sonya Bangmann. New York: Crown.

Evans, David W., trans. (1992) *Discourses of Gotama Buddha: Middle Collection.* London: Janus.

Faison, Seth. (1999) "In the Dalai Lama's Homeland, Tibetans Are Taught Atheism." *New York Times International,* February 4, 1999.

Feyerabend, Paul. (1994a) "Has the Scientific View of the World a Special Status Compared with Other Views?" In *Physics and Our View of the World,* edited by Jan Hilgevoord. Cambridge: Cambridge University Press. Pp. 135–45.

———. (1994b) "Quantum Theory and Our View of the World." In *Physics and Our View of the World,* edited by Jan Hilgevoord. Cambridge: Cambridge University Press. Pp. 149–68.

Feynman, Richard. (1983) *The Character of Physical Law.* Cambridge, Mass.: MIT Press.

Feynman, Richard, R. B. Leighton, and M. Sands. (1963) *The Feynman Lectures on Physics.* Reading, Mass.: Addison-Wesley.

Flavell, J. H., and H. M. Wellman. (1977) "Metamemory." In *Perspectives on the Development of Memory and Cognition,* edited by R. V. Kail, Jr., and J. W. Hagen. Hillsdale, N.J.: Erlbaum.

Foster, Michael. (1934) "The Christian Doctrine of Creation and the rise of Modern Natural Science," *Mind,* 43, pp. 446–68.

Forman, Robert K. C. (1990a) "Mysticism, Constructivism, and Forgetting." In *The Problem of Pure Consciousness.* New York: Oxford University Press. Pp. 3–49.

———. (1990b) "Eckhart, Gezücken, and the Ground of the Soul." In *The Problem of Pure Consciousness.* New York: Oxford University Press. Pp. 98–120.

———. (1994) " 'Of Capsules and Carts': Mysticism, Language and the Via

Negativa." *Journal of Consciousness Studies: Controversies in Science and the Humanities*, 1, no. 1, pp. 38–49.

Foster, Michael. (1934) "The Christian Doctrine of Creation and the Rise of Modern Natural Science." *Mind*, 43, pp. 446–68.

———. (1936) "Christian Theology and Modern Science of Nature." *Mind*, 45, pp. 1–28.

Funkenstein, Amos. (1986) *Theology and the Scientific Imagination*. Princeton: Princeton University Press.

Galin, David. (1992) "Theoretical Reflections on Awareness, Monitoring, and Self in Relation to Anosognosia," *Consciousness and Cognition*, 1, pp. 152–62.

Galison, Peter, and David J. Stump, eds. (1996) *The Disunity of Science: Boundaries, Contexts, and Power*. Stanford, Calif.: Stanford University Press.

Gardner, Howard. (1985) *The Mind's New Science*. New York: Basic Books.

Ginsburg, H., and S. Opper. (1979) *Piaget's Theory of Intellectual Development*. Englewood Cliff, N.J.: Prentice-Hall.

Goleman, Daniel. (1995) *Emotional Intelligence*. New York: Bantam.

———, ed. (1997) *Healing Emotions: Conversations with the Dalai Lama on Mindfulness, Emotions, and Health*. Boston: Shambhala.

Gooding, David, and Frank James, eds. (1985) *Faraday Rediscovered*. New York: Stockton Press.

Gribbin, John. (1984) *In Search of Schrödinger's Cat: Quantum Physics and Reality*. New York: Bantam Books.

Griffin, David Ray. (1989) *God and Religion in the Postmodern World*. Albany: State University of New York Press.

Grush, Rick, and Patricia S. Churchland. (1995) "Gaps in Penrose's Toilings." in *Journal of Consciousness Studies: Controversies in Science and the Humanities*, 2, no. 1, pp. 10–29.

Güzeldere, Güven. (1995) "Consciousness: What It Is, How To Study It, What to Learn from Its History." *Journal of Consciousness Studies: Controversies in Science and the Humanities*, 2, no. 1, pp. 30–51.

———. (1998) "The Many Faces of Consciousness: A Field Guide." In *The Nature of Consciousness: Philosophical Debates*, edited by N. Block, O. Flanagan, and G. Güzeldere. Cambridge, Mass.: MIT Press. Pp. 1–67.

Gyaltrul Rinpoche. (1993) *Ancient Wisdom: Nyingma Teachings on Dream Yoga, Meditation, and Transformation*. Ithaca, N.Y.: Snow Lion.

Hacking, Ian. (1983) *Representing and Intervening*. Cambridge: Cambridge University Press.

Hans, James S. (1993) *The Mysteries of Attention*. Albany: State University of New York Press.

Hardin, G. (1988) *Color for Philosophers: Unweaving the Rainbow*. Indianapolis: Hackett.

Harrison, Edward R. (1981) *Cosmology: The Science of the Universe*. New York: Cambridge University Press.

Harvey, Peter. (1995) *The Selfless Mind*. Surrey, England: Curzon.

Hayward, Jeremy W., and Francisco J. Varela, eds. (1992) *Gentle Bridges: Conversations with the Dalai Lama on the Sciences of Mind*. Boston: Shambhala.

Heisenberg, Werner. (1962) *Physics and Philosophy: The Revolution in Modern Science*. New York: Harper and Row.

————. (1971) *Physics and Beyond: Encounters and Conversations*. New York: Harper and Row.

Herbert, Nick. (1985) *Quantum Reality: Beyond the New Physics*. Garden City, N.Y.: Anchor Books.

Hesse, Mary B. (1994) "The Source of Models for God: Metaphysics or Metaphor?" In *Physics and Our View of the World*, edited by Jan Hilgevoord. Cambridge: Cambridge University Press. Pp. 239–54.

Hilgevoord, Jan, ed. (1994) *Physics and Our View of the World*. Cambridge: Cambridge University Press.

Hodgson, David. (1994) "Neuroscience and Folk Psychology: An Overview." *Journal of Consciousness Studies: Controversies in Science and the Humanities*, 2, no. 2, pp. 205–16.

Hodson, Phyllis, ed. (1955) *Deonise Hid Diuinite*. London: Oxford University Press.

Hofstadter, Douglas R., and Daniel Dennett, eds. (1981) *The Mind's Eye: Fantasies and Reflections on Self and Soul*. New York: Basic Books.

Horgan, John. (1996) *The End of Science: Facing the Limits of Knowledge in the Twilight of the Scientific Age*. Reading, Mass.: Addison-Wesley.

Horgan, Terrence, and James Woodward. (1990) "Folk Psychology Is Here to Stay." In *Mind and Cognition: A Reader*, edited by William G. Lycan. Cambridge, Mass.: Blackwell. Pp. 399–420.

Horton, Robin. (1982) "Tradition and Modernity Revisited." In *Rationality and Relativism*, edited by Martin Hollis and Steven Lukes. Cambridge, Mass.: MIT Press. Pp. 201–60.

Houshmand, Zara, Robert Livingston, and B. Alan Wallace, eds. (1999) *Consciousness at the Crossroads: Conversations with the Dalai Lama on Brainscience and Buddhism*. Ithaca, N.Y.: Snow Lion.

Hume, David. (1980) *Dialogues concerning Natural Religion*. Edited by Richard H. Popkin. Indianapolis: Hackett.

Indich, William M. (1995) *Consciousness in Advaita Vedanta*. Delhi: Motilal Barnarsidass.

Jackendoff, Ray. (1987) *Consciousness and the Computational Mind*. Cambridge, Mass.: MIT Press.

Jackson, Frank. (1990) "Epiphenomenal Qualia." In *Mind and Cognition: A Reader*, edited by William G. Lycan. Cambridge, Mass.: Basil Blackwell. Pp. 469–77.

James, William. (1890/1950) *The Principles of Psychology*. New York: Dover.

————. (1892) "A Plea for Psychology as a Science." *Philosophical Review*, 1, pp. 146–53.

————. (1897/1979) *The Will to Believe and Other Essays*. Cambridge, Mass.: Harvard University Press.

————. (1899/1958) *Talks to Teachers: On Psychology; and to Students on Some of Life's Ideals*. New York: Norton.

————. (1902/1960) *The Varieties of Religious Experience*. Huntington, N.Y.: Fontana.

————. (1902/1985) *The Varieties of Religious Experience: A Study in Human Nature*. New York: Penguin.

————. (1904/1977) "Does Consciousness Exist?" In *The Writings of William James*, edited by John J. McDermott. Chicago: University of Chicago Press. Pp. 169–83.

————. (1905/1977) "The Notion of Consciousness." In *The Writings of William James*, edited by John J. McDermott. Chicago: University of Chicago Press. Pp. 184–94.

————. (1907/1975) *Pragmatism*. Cambridge, Mass.: Harvard University Press.

————. (1909/1977) *A Pluralistic Universe*. Cambridge, Mass.: Harvard University Press.

————. (1911) *Some Problems of Philosophy: A Beginning of an Introduction to Philosophy*. London: Longmans, Green.

————. (1911/1977) "Faith and the Right to Believe." In *The Writings of William James*, edited by John J. McDermott. Chicago: University of Chicago Press. Pp. 735–40.

————. (1912) *The Will to Believe and Other Essays in Popular Philosophy*. New York: Holt.

————. (1912/1947) *Essays in Radical Empiricism*. New York: Longmans, Green.

————. (1912/1976) *Essays in Radical Empiricism*. Cambridge, Mass.: Harvard University Press.

————. (1989) *Essays in Religion and Morality*. Cambridge, Mass.: Harvard University Press.

Johnson, Michael G., and Tracy B. Henley. (1990) *Reflections on* The Principles of Psychology: *William James after a Century*. Hillsdale, N.J.: Erlbaum.

Kail, R. V., Jr., and J. W. Hagen, eds. (1977) *Perspectives on the Development of Memory and Cognition*. Hillsdale, N.J.: Erlbaum.

Kalupahana, David J. (1987) *The Principles of Buddhist Psychology*. Albany: State University of New York Press.

Kamenetz, Rodger. (1994) *The Jew in the Lotus: A Poet's Rediscovery of Jewish Identity in Buddhist India*. San Francisco: HarperSanFrancisco.

Kant, Immanuel. (1781/1998) *Critique of Pure Reason*. Translated by Paul Guyer and edited by Allen W. Wood. New York : Cambridge University Press.

————. (1786/1970) *Metaphysical Foundations of Natural Science*. Translated by James Ellington. Indianapolis: Bobbs-Merrill.

————. (1788/1997) *Critique of Practical Reason*. Translated and edited by Mary Gregor. New York: Cambridge University Press, 1997.

Karma Chagmé. (1998) *A Spacious Path to Freedom: Practical Instructions on the Union of Mahāmudrā and Atiyoga*, with commentary by Gyatrul Rinpoche. Translated by B. Alan Wallace. Ithaca, N.Y.: Snow Lion.

Katz, Steven T. (1978) "Language, Epistemology, and Mysticism." In *Mysticism and Philosophical Analysis*, edited by Steven T. Katz. New York: Oxford University Press. Pp. 22–74.

————. (1983) "The 'Conservative' Character of Mystical Experience" in *Mysticism and Religious Traditions*, edited by Steven T. Katz. Oxford: Oxford University Press. Pp. 3–60.

Khenchen Kunzang Palden and Minyak Kunzang Sönam. (1993) *The Nectar of Mañjuśrī's Speech*. In *Wisdom: Two Buddhist Commentaries*. Peyzac-le-Moustier, France: Padmakara.

Koestler, Arthur. (1959) *The Sleepwalkers: A History of Man's Changing Vision of the Universe*. New York: Macmillan.

———. (1967) *The Ghost in the Machine*. New York: Macmillan.

Knight, Margaret, ed. (1950) *William James: A Selection from His Writings on Psychology*. Harmondsworth, England: Penguin Books.

Kuhn, Thomas S. (1970) *The Structure of Scientific Revolutions*. 2nd ed. Chicago: University of Chicago Press.

Lamrimpa, Gen. (1995) *Calming the Mind: Tibetan Buddhist Teachings on the Cultivation of Meditative Quiescence*. Translated by B. Alan Wallace. Ithaca, N.Y.: Snow Lion.

———. (1999) *Realizing Emptiness: The Madhyamaka Cultivation of Insight*. Ithaca, N.Y.: Snow Lion.

Lashley, K. (1956) "Cerebral Organization and Behavior" in *The Brain and Human Behavior*, edited by H. Solomon, S. Cobb, and W. Penfield. Baltimore: Williams and Wilkins Press. Pp. 1–18.

Lemonick, Michael D. (1995) "Glimpses of the Mind: What Is Consciousness? Memory? Emotion? Science Unravels the Best-kept Secrets of the Human Brain." *Time*, July 17, 1995.

Levinson, Henry Samuel. (1981) *The Religious Investigations of William James*. Chapel Hill: University of North Carolina Press.

Liu Kiufeng. (1996) *Mathematics and Science Curriculum Change in the People's Republic of China*. Lewiston, N.Y.: Edwin Mellen Press.

Lowe, E. J. (1996) *Subjects of Experience*. Cambridge: Cambridge University Press.

Loy, David. (1988) *Nonduality: A Study in Comparative Philosophy*. New Haven: Yale University Press.

Lukes, Steven. (1973) *Emile Durkheim: His Life and Work, A Historical and Critical Study*. London: Penguin Press.

Lycan, William G., ed. (1990) *Mind and Cognition: A Reader*. Cambridge, Mass.: Blackwell.

Lyons, William. (1986) *The Disappearance of Introspection*. Cambridge, Mass.: MIT Press.

Mackworth, J. F. (1970) *Vigilance and Attention*. New York: Penguin Books.

Mackworth, N. H. (1950) Medical Research Council Special Report no. 268. London: H. M. Stationary Office.

Malinowski, Bronislaw. (1925/1948) *Magic, Science and Religion and Other Essays*. Boston: Beacon Press.

Marcel, A. J., and E. Bisiach, eds. (1992) *Consciousness in Contemporary Science*. Oxford: Oxford University Press.

Markman, E. M. (1977) "Realizing What You Don't Understand: A Preliminary Investigation." *Child Development*, 48, pp. 986–92.

Matt, Daniel C. (1995) *The Essential Kabbalah: The Heart of Jewish Mysticism*. San Francisco: HarperSanFrancisco.

Mauss, Marcel. (1972) *A General Theory of Magic*. Translated by Robert Brain. New York: Norton.

McCann, Justin, ed. (1952) *The Cloud of Unknowing and Other Treatises*. Westminster, Md.: Newman Press.

McDermott, John J., ed. (1977) *The Writings of William James: A Comprehensive Edition*. Chicago: University of Chicago Press.

McFarlane, Thomas. (1996) "Integral Science: Toward a Comprehensive Science of Inner and Outer Experience." Unpublished manuscript. http://www.integralscience.org/tom

———. (1997) "Relativity: Inside and Out" Unpublished manuscript.http://www.integralscience.org/tom

McMullin, Ernan. (1994) "Enlarging the Known World." In *Physics and Our View of the World*, edited by Jan Hilgevoord. Cambridge: Cambridge University Press. Pp. 79–113.

Medawar, Peter. (1984) *Pluto's Republic*. Oxford: Oxford University Press.

Merrell-Wolff, Franklin. (1994) *Experience and Philosophy*. Albany: State University of New York Press.

———. (1995) *Transformations in Consciousness*. Albany: State University of New York Press.

Merton, Robert K. (1938) *Science, Technology and Society in Seventeenth-Century England*. New York: Howard Fertig.

———. (1949) *Social Theory and Social Structure*. Glencoe, Ill.: Free Press of Glencoe.

Monod, Jacques. (1971) *Chance and Necessity*. New York: Knopf.

Myers, Gerald E. (1986) *William James: His Life and Thought*. New Haven: Yale University Press.

Nagel, Thomas. (1986) *The View from Nowhere*. New York: Oxford University Press.

Nichols, Susan. (1999) "Alternative Interventions for Children with ADD." Unpublished essay.

Oakley, Francis. (1961) "Christian Theology and the Newtonian Science: The Rise of the Concept of the Laws of Nature" *Church History*, 30, pp. 433–57.

Padmasambhava. (1998) *Natural Liberation: Padmasambhava's Teachings on the Six Bardos*. With a commentary by Gyatrul Rinpoche. Translated by B. Alan Wallace. Boston: Wisdom.

Pickering, Andrew. (1984) *Constructing Quarks*. Chicago: University of Chicago Press.

Pieper, Josef. (1966) *Happiness and Contemplation*. Translated by Richard and Clara Winston. Chicago: Henry Regnery.

Physicians' Desk Reference. (1995) 49th ed. Montvale, N.J.: Medical Economics Data.

Polanyi, Michael. (1966) *The Tacit Dimension*. New York: Doubleday.

Posner, M. I. (1978) *Chronometric Exploration of Mind*. Hillsdale, N.J.: Lawrence Erlbaum Associates.

Prabhavananda, Swami, and Christopher Isherwood, trans. (1981) *How to Know God: The Yoga Aphorisms of Patanjali*. Hollywood, Calif.: Vedanta Press.

Putnam, Hilary. (1975) "Philosophy and Our mental Life." In *Mind, Language and Reality*. Cambridge: Cambridge University Press. Pp. 291–303.

———. (1987) *The Many Faces of Realism*. La Salle, Ill.: Open Court.

————. (1988) *Representation and Reality*. Cambridge, Mass.: MIT Press.

————. (1990) *Realism with a Human Face*. Edited by James Conant. Cambridge, Mass.: Harvard University Press.

————. (1991) "Replies and Comments." *Erkenntnis*, 34, no. 3, pp. 401–424.

————. (1997) "James's Theory of Truth." In *The Cambridge Campanion to William James*, edited by Ruth Anna Putnam. Cambrige: Cambridge University Press.

Putnam, Ruth Anna, ed. (1997) *The Cambridge Companion to William James*. Cambrige: Cambridge University Press.

Raju, P. T. (1985) *Structural Depths of Indian Thought*. Albany: State University of New York Press.

Remy, Lorraine Nicolas. (1595/1930) *Demonolatry*. Translated by E. A. Ashwin. London: J. Rodker.

Rolston, Holmes, III. (1978) *Science and Religion: A Critical Survey*. New York: Random House.

Rothberg, Donald. (1990) "Contemporary Epistemology and the Study of Mysticism." In *The Problem of Pure Consciousness Mysticism and Philosophy*, edited by Robert K. C. Forman. New York: Oxford University Press. Pp. 163–210.

Russell, Bertrand. (1961). *Religion and Science*. New York: Oxford University Press.

Sanders, A. F., ed. (1970) *Attention and Performance*. Vol. 1. Amsterdam: North-Holland.

Schleiermacher, Friedrich. (1799/1988) *On Religion: Speeches to Its Cultured Despisers*. Translated by Richard Crouter. New York: Cambridge University Press.

Schlipp, Paul Arthur, ed. (1969) *Albert Einstein: Philosopher-Scientist*. Vol. 1. La Salle, Ill.: Open Court.

Searle, John R. (1994) *The Rediscovery of the Mind*. Cambridge, Mass.: MIT Press.

Sellars, Wilfred. (1963) *Science, Perception, and Reality*. Atlantic Highlands, N.J.: Humanities Press.

Shallice, T. (1972) "Dual Functions of Consciousness." *Psychological Review*, 79, 5, pp. 383–93.

Shapere, Dudley (1982) "The Concept of Observation in Science and Philosophy." *Philosophy of Science*, 49, No. 4, pp. 485–525.

Shapin, Steven, and Simon Schaffer. (1985) *Leviathan and the Air Pump*. Princeton: Princeton University Press.

Shorto, R. (1997) "Belief by the Numbers." *New York Times Magazine*, January 7, 1997.

Skinner, B. F. (1974) *About Behaviorism*. New York: Knopf.

Sleator, Esther K., and William E. Pelham, Jr. (1986) *Attention Deficit Disorder*. Norwalk, Conn.: Appleton-Century-Crofts.

Sloan, Douglas. (1983) *Insight-Imagination: the Emancipation of Thought and the Modern World*. Westport, Conn.: Greenwood Press.

Smith, Huston. (1982) *Beyond the Post-Modern Mind*. New York: Crossroad Press.

Smith, Peter, and O. R. Jones. (1988) *The Philosophy of Mind: An Introduction*. Cambridge: Cambridge University Press.

Sprat, Thomas. (1667/1959) *The History of the Royal Society of London*, edited by J. I. Cape and H. W. Jones. London: Routledge.

Squires, Euan J. (1990) *Conscious Mind in the Physical World*. Bristol England: Adam Hilger.

―――. (1994) "Quantum Theory and the Need for Consciousness." *Journal of Consciousness Studies: Controversies in Science and the Humanities*, 1, no. 2, pp. 201–204.

Stapp, Henry P. (1993) *Mind, Matter and Quantum Mechanics*. Berlin: Springer-Verlag.

Stich, Stephen. (1983) *From Folk Psychology to Cognitive Science: The Case against Belief*. Cambridge, Mass.: Bradford.

Sutherland, Keith. (1994) "Consciousness—Its Place in Contemporary Science." *Journal of Consciousness Studies: Controversies in Science and the Humanities*, 1, no. 2, pp. 285–86.

Sutherland, N. S., ed. (1989) *The International Dictionary of Psychology*. New York: Continuum.

Tallis, Raymond. (1994) *Psycho-electronics*. London: Ferrington.

Tambiah, Stanley J. (1990) *Magic, Science, Religion and the Scope of Rationality*. Cambridge: Cambridge University Press.

Taylor, Charles. (1989) *Sources of the Self: The Making of the Modern Identity*. Cambridge, Mass.: Harvard University Press.

Taylor, Eugene. (1990) "New Light on the Origin of William James's Experimental Psychology." In *Reflections on* The Principles of Psychology: *William James after a Century*, edited by Michael G. Johnson, and Tracy B. Henley. Hillsdale, N.J.: Erlbaum. Pp.–.

Thomas, Keith. (1971) *Religion and the Decline of Magic*. New York: Scribner's.

Thompson, Evan. (1995) *Colour Vision: A Study in Cognitive Science and the Philosophy of Perception*. New York: Routledge.

Thrangu Rinpoche. (1993) *Buddha Nature*. Hong Kong: Rangjung Yeshi.

Vajirañāṇa, Paravahera. (1975) *Buddhist Meditation in Theory and Practice*. Kuala Lumpur: Buddhist Missionary Society.

Van Fraassen, Bas C. (1989) *The Scientific Image*. Oxford: Clarendon Press.

Varela, Francisco, Evan Thompson, and Eleanor Rosch. (1991) *The Embodied Mind: Cognitive Science and Human Experience*. Cambridge, Mass.: MIT Press.

―――, ed. (1997) *Sleeping, Dreaming, and Dying: An Exploration of Consciousness with the Dalai Lama*. Boston: Wisdom.

Vasubandhu. (1991) *Abhidharmakośabhāṣyam*. French translation by Louis de La Vallée Poussin. English translation by Leo M. Pruden. Berkeley, Calif.: Asian Humanities Press.

von Aster, E. (1908) "Die psychologische Beobachtung und experimentelle Untersuchung von Denkvorgängen." *Zeitschrift für Psychologie*, 49 (1908), pp. 56–107.

von Gierke, Otto. (1927) *Political Theories of the Middle Ages*. Translated by F. W. Maitland. Cambridge: Cambridge University Press.

Wallace, B. Alan. (1996) *Choosing Reality: A Buddhist View of Physics and the Mind*. Ithaca, N.Y.: Snow Lion.

————. (1998) *The Bridge of Quiescence: Experiencing Tibetan Buddhist Meditation*. Chicago: Open Court.

————. (1999) "The Buddhist Tradition of *Śamatha*: Methods for Refining and Examining Consciousness." *Journal of Consciousness Studies: Controversies in Science and the Humanities*, 6, no. 2–3, pp. 175–87.

Walshe, M. O'C, trans. (1979) *Meister Eckhart: German Sermons and Treatises*. 2 vols. London: Watkins.

Watson, John. (1913) "Psychology as the Behaviorist Views It." *Psychological Review*, 20, pp. 158–77.

————. (1913/1970) *Behaviorism*. New York: Norton.

Weber, Renée. (1986) *Dialogues With Scientists and Sages*. New York: Routledge and Kegan Paul.

Whitehead, Alfred North. (1938) *Modes of Thought*. New York: Macmillan.

————. (1975) *Science and the Modern World: Lowell Lectures, 1925*. New York: Fontana Books.

Whitehead, W. F., ed. (1897/1971) *The Three Books of Occult Philosophy*. Aquarian Press.

Whittaker, E. T. (1954) *A History of the Theories of Aether and Electricity, 1900–1926*. New York: Philosophical Library.

Wiener, N. (1961) *Cybernetics, or Control and Communication in the Animal and the Machine*. 2nd ed. Cambridge, Mass.: MIT Press (1st ed. 1948).

Wigner, Eugene. P. (1967) *Symmetries and Reflections: Scientific Essays of Eugene P. Wigner*. Bloomington: Indiana University Press.

Wilkes, Kathleen. (1988) "—Yishi, Duh, Um and Consciousness." In *Consciousness in Contemporary Science*, edited by A. J. Marcel and E. E. Bisiach. Oxford: Oxford University Press, 1992. Pp. 16–41.

Williams, Paul. (1989) *Mahāyāna Buddhism: The Doctrinal Foundations*. London: Routledge.

Wilson, D. (1983) *Rutherford: Simple Genius*. Cambridge, Mass.: MIT Press.

Wilson, Edward O. (1998a) "The Biological Basis of Morality." *Atlantic Monthly*, April 1998, pp. 53–70.

————. (1998b) *Consilience: The Unity of Knowledge*. New York: Knopf.

Woodworth, Robert, and Harold Schlosberg. (1955) *Experimental Psychology*. 3d ed. London: Methuen.

Zajonc, Arthur. (1993) *Catching the Light: The Entwined History of Light and Mind*. New York: Bantam Books.

Zaner, R. (1988) *Death: Beyond the Whole-Brain Criteria*. Boston: Kluwer Academic. Pp. 147–170.

INDEX

Absolutes, historical progression
toward relativity and away
from, 32, 72
Anderson, John, and scientific
materialist view of mind, 60,
148–150
Animism, anthropomorphic, 124–
125
Antirealism (scientific), 19, 20, 21,
66
Aquinas, Thomas, on need for
professional contemplatives,
187
Aristotle, 42, 48
Asaṅga, and Mahāyāna Buddhist
contemplative tradition, 105,
106–108, 185
Asymmetry
between consciousness and matter
in James, 114
between consciousness and objects
of consciousness, and in mind/
body interaction, 72
between the emergent
phenomenon and that which
gives rise to it, 134
as sign of degenerate theory, 72

Attention. *See also* Great Perfection
tradition; Introspection; James,
William
development of, in Mahāyāna
Buddhism, 103–109
and fullness of life, 101
neuroscientific studies of, 99–100
Attention deficit disorders (ADD),
100–101
Augustine, 42, 76, 77, 103, 186–187
contemplative training of, 101–102
and mental perception, 76, 91
and origin of human soul, 4, 129–
130

Bacon, Francis, experimental
philosophy of, 48
Behaviorism, 28–29, 78, 124
Believing is seeing, 60–62, 125
Bell, John, on reality and quantum
theory, 69
Boorstin, Daniel J., 62, 179
Boutroux, Émile, on continuity
between subjective and
objective, 79
Boyle, Robert, 34, 49, 51–52
Buddhaghosa, 103–105

Buddhism, Mahāyāna, cultivation of attention and nature of mind in, 105–109

Buddhism, Theravāda, cultivation of attention and nature of mind in, 103–105

Buddhism, Tibetan, 9–10, 109–112, 115–118, 119

Cārvāka, early Indian materialism of, 161–162

Causality, 69, 70–71, 81–82, 137–138, 142–143

China, materialist ideology of, 10

Christian, William, on religion and religious discourse, 171–172

Christianity. *See also* Christian theology; Contemplative practice; God; Protestantism

and consciousness after death, 5, 6–7

contemporary unbelief in aspects of, 55

and creation ex nihilo, 42

decline of contemplative practice in, 102

magic and miracles in medieval, 43–47,

on origins of consciousness, 4

Christian theology

scientific materialism's defense of, 49–52

and theory of imposed natural law, 50–51

and true miracles versus mere marvels, 50

Churchland, Patricia, on eliminative materialism, 26–27, 128, 161

Churchland, Paul, on eliminative materialism, 29, 138–139, 161

Closure principle, 12, 24–25

in early mechanical/experimental philosophy, 53

incompatibility of experience and, 25

and lack of causal efficacy of mental phenomena, 81

and neglect of will in modern psychology, 98

possible violation of, in quantum mechanics, 142–143

Cognition(s), 92–93, 95, 124–126, 131–132, 148–149. *See also* Cognitive psychology; Cognitive science(s); Consciousness; Introspection; Mind

Cognitive psychology, 27, 124, 148–150. *See also* Cognitive science(s)

Cognitive science(s), 80–81, 168. *See also* Consciousness; Mind; Neuroscience(s); Scientific materialism

and redefinition of subjective cognitive terms, 65, 124–126, 148–150

Comte, Auguste, on scientific positivism, 37

Consciousness. *See also* Introspection; James, William; Mind; Qualia; Radical Empiricism

author's usage of the term, 5–6

and contemplative practice, 103–112, 115–120, 178–181, 182–183

field versus atomistic theories of, 114

as an object of consciousness, 92–93, 109–112

problems with scientific understanding of, 3–4, 5, 145, 150, 178

pure, 116–120

scientific materialism and theories of, 3–4, 27–28, 146–147, 148–150, 150–158

Conservation principle, 25, 35–36, 142–143

Contemplative practice. *See also* Buddhism; Language

in Christianity, 101–102

and conceptual meditation, 108, 109, 116–117, 118–119, 173, 183

and contemplative experience as
    gold reserve for religious texts,
    187
and guidelines for scientific study
    of contemplative experience,
    180–181
and hypothesis of joy arising from
    nature of consciousness, 108
as interface between science and
    religion, 120, 178–184, 187–
    188
role of experienced practitioners in
    both science and, 183
Copernicus, 47
Crick, Francis, on consciousness, 128,
    146, 147

Dalai Lama, 9, 10, 11, 165, 166
Deism, 26
Dennett, Daniel, 76, 139, 158
Dependently related events
    mental and material phenomena
        as, 68, 75, 135
    subjective and objective as, 134
    universe as consisting solely of, 71–
        72
Descartes, René, 25, 51, 72, 185
    cosmography of, 47–48
    introspection in philosophy of, 76,
        93–95
    mechanical philosophy of, and
        roots of scientific materialism,
        47–48
    and physical world as domain of
        natural philosophy and science,
        25, 59
Devil, in medieval Christianity and
    theology, 43–46
Discovery, verifiability of
    contemplative and scientific,
    184
Dogma. See also Fundamentalism
    author's usage of the term, 22
    in medieval scholasticism and
        Scientific Revolution, 21–22
    in religion and science, 22, 129–
        130

Durkheim, Emile
    and mana as precursor to scientific
        concept of energy, 35
    and science as higher form of
        religion, 36
    theory of religion of, 33–37

Eckhart, Meister, on contemplative
    practice, 101–102, 103
Einstein, Albert, 26, 35, 37, 83
Eliminative materialism, 29, 138–141,
    161. See also Consciousness;
    Mind; Scientific materialism
Emergent properties, analysis of
    nature of, 135–137
Emotional intelligence, and self-
    monitoring, 168
Energy. See also Conservation
    principle
    contemporary scientific ignorance
        regarding nature of, 36, 134
    as mana, 35–36
Environmental problems, relation of
    internal human problems to, 9
Experience. See also James, William,
    and under Contemplative
    practice; Radical Empiricism
    precognitive structuring of, 118
    in sixteenth-century organic
        philosophies, 44
Experimental philosophy, in
    mechanical and organic views
    in sixteenth century, 44. See
    also Mechanical/experimental
    philosophy

Fact(s), empirical
    and metaphysically loaded
        interpretation of fact, 18
    verifiability of, by first-person and
        third-person means, 22
Faith
    in contemplative practice and in
        science, 175, 184
    in future of neuroscience, 86, 140,
        144, 153
    in religion and science, 157,186–187

Faith (*continued*)
  in science, 18, 174–175
  and scientific materialism, 31–32,
    140, 144, 157
Fetus, Christian and scientific views
  on consciousness of, 4
Feynman, Richard, on energy, 35–36,
  134
Folk psychology, 85, 86, 91
  problems with eliminative
    materialist theory of, 139–140
Forman, Robert K. C., on pure
  consciousness, 118
Fraassen, Bas van, antirealism of, 20,
  66
Free will, in science and Christianity,
  4–5
Fundamentalism, in religion and
  science, 38–39, 186

Galileo, 20, 32, 35
Gardner, Howard, on cognitive
  psychology and cognitive
  science, 125–126
God
  incompatibility of science and a
    personal, 26
  role of, in nature, 49–55
God-of-the-gaps, 54–55, 128, 147
Great Perfection tradition
  (Dzogchen), 109–112, 115–118,
  119
Güzeldere, Güven, 26, 132–133

Hacking, Ian, 20, 181
Hallucination, self-induced visual, 60–
  61
Hans, James, on attention, 101
Happiness, human, 9, 19, 101, 159–
  161, 163–164
Heisenberg, Werner, on quantum
  theory, 71, 83, 134
  Uncertainty Principle of, 80
Herbert, Nick, on quantum
  paradoxes, 120
Hippocrates, and brain as source of
  mental phenomena, 42

Horgan, John, and "ironic science,"
  144
Hume, David, and definition in
  terms of function, 135
Hypothesis, 18, 183

Ideology, and science, 10, 139
Illusions of knowledge, as obstacles
  to discovery, 62
Imagination, human (*vis imaginativa*),
  in Renaissance organic
  philosophy, 44–45
Information
  computer modeling of human
    processing of, 148–149
  and sanitization of the term, 124–
    126
Inquiry
  faith in power of human, 46
  scientific, in early days of
    Scientific Revolution and at
    present, 163
  scientific, as form of worship and
    sacred activity, 34, 51–52
  scientific and contemplative, 182–
    183, 184
Intentionality, 130–131, 151
Introspection, 75–96
  arguments against, 79–83, 83–85,
    86–87
  in Augustine and later Christian
    contemplatives, 102
  double taboo of, 75
  and failure of early
    introspectionist school, 89
  fallibility of, 93–96
  as folk psychology and culturally
    conditioned, 85–86
  historical sketch of, 76–79
  improving and atrophying of
    powers of, 82–83
  neglect of, in modern psychology
    and brain sciences, 75–76
  problems with, 78, 84–85, 94–95
  a re-evaluation of, 89–96
  and refining of attention, 90, 96
  as retrospection, 87, 92

and self-monitoring, 168
training in, as analogous to
training in scientific
observation, 86
twentieth-century failure to
understand, 78–79
Invariance. *See also* Perception
of laws of physics and
mathematics, and in certain
religious statements, 67

James, William, 59–60, 95–96. *See
also* Radical Empiricism
attentional reality principle of, 59,
78, 98, 138
on difficulty of focusing attention
on nameless, 91
and early psychology in United
States, 59–60
on faith as an essential attitude,
186
and introspection, 88–89, 92, 94,
and many worlds of consciousness,
188
on nature of consciousness, 112–
114
proposal for a science of religion
of, 177, 179–180
on sustained, voluntary attention,
97–100
and three theories of correlation
between mental and brain
states, 129, 138
Josephson, Brian, on pure
consciousness, 120

Katz, Steven, on contemplative
experience, 173–174, 180, 182
Koch, Christof, on consciousness,
128, 147
Kuhn, Thomas, and nonscientific
influences on science, 22

Language
descriptive and evocative, and
scientific and contemplative
uses of, 182–183

necessity of, in all descriptions of
world, 68
and truth, 64–65
Leibniz, G. W., 48, 51, 55, 56, 80
Lemonick, Michael, on brain and
consciousness, 60, 146–147
Lyons, William, and argument
against introspection, 85–87

Magic. *See under* Christianity
Marxism, spread of scientific
materialism, religious
suppression and, 165–166
Materialism, and atheism as logical
conclusions of self-contained,
self-sufficient universe, 55
Mathematics, privileged role of, 35,
42
Matter, in contemporary physics and
quantum theory, 69–72, 134,
142–143
Mechanical/experimental philosophy,
52–54. *See also* Mechanical
philosophy
Mechanical philosophy
as infant stage of contemporary
scientific materialism, 50, 51
and nature as divorced from
perceptual experience, 123
and suppression of first-person,
phenomenological modes of
inquiry, 49–50
triumph of, over natural magic, 46
use of, as a defense against heresy,
49
Medicine, modern, ideological
domination of scientific
materialism in, 28, 167–168
Mental perception(s), 91–93. *See also*
Consciousness; Contemplative
practice; James, William; Mind;
Searle, John
in Augustine, 76
and conceptual superimposition,
118–119
fallibility of, 97

Mental perceptions(s) (*continued*)
  as sole means of observing mental
    phenomena, 97
  validity of all, with respect to
    appearances of mental
    phenomena, 95
Mental phenomena. *See*
    Consciousness; Contemplative
    practice; James, William;
    Mental perception(s); Mind;
    Qualia
Metacognition, 85–87, 108. *See also*
    Introspection
Mind. *See also* Consciousness; Qualia;
    Radical Empiricism
  causal agency of, 141–142
  and consciousness as central to
    both religion and science, 6,
    172, 188
  and distinction between brain states
    and mental states, 131–132
  as an emergent property of brain,
    133–138
  failure of science to understand
    and control, 12, 18–19
  as fundamental instrument of
    scientific inquiry, 29
  identity of states of brain and, 127–
    132
  information-processing approach
    to study of, 124–126, 148–150
  as irreducibly first-person
    phenomenon, 150, 151
  marginalization of, 27–30
  and mental events as accessible
    only by first-person means, 22
  and mental events as unconscious
    and devoid of causal efficacy,
    80–81
  and mental representations being
    perceived versus identifed as,
    95
  and mind/body problem, 70, 72,
    81, 132, 137, 143, 148
  as a nonentity, 138–141
Mind and Life conferences, 11
Mind/body problem. *See under* Mind
Miracles. *See under* Christianity

Monism, 12, 23
  in early Greek thought, 41–42
  and mass/energy as fundamental
    stuff of universe, 23
  as a mere hypothesis, 60
Monod, Jacques, on objectivity as an
    ethical choice, 22
Mutual contingency, in causal,
    spatial, temporal, and cognitive
    relations, 71

Neuromythology, 145, 149
Neuroscience(s)
  and consciousness after death, 5, 6–
    7
  and definition of consciousness in
    terms of neurophysiological
    correlates, 132
  dominance of scientific realism
    among philosophers of, 20
  and exclusion of subjectivity, 153
  and ignorance regarding
    consciousness and functioning
    of brain, 90, 128
  and lack of one-to-one
    relationship between
    wavelength and perceived
    color of light, 127
  and reduction of mental events to
    brain activity, 5, 24
  reliance on subjective accounts of
    mental phenomena in, 90
Newton, Isaac
  ambivalence of, regarding
    (ontological) hypotheses, 34–35
  on consequences of a self-
    contained, self-sufficient
    universe, 55
  interpretation of spiritual entities
    of, 53
  and the rise of science, 20, 32, 49,
    50, 51, 52
Noetic revolution, need for and
    potentials of, 13, 56
Nonduality, 63–64, 117. *See also*
    Contemplative practice
N-rays, and "believing is seeing," 60–
    61

Objectivism, 12, 22–23, 35, 41, 47
    challenge to, 64
    and exclusion of subjective
        elements of human mind, 29–
        30
    incompatibility of study of
        subjective, introspective
        phenomena and, 22–23, 79
    original theological underpinnings
        of, 26
Observation
    as interaction, 29–30
    theory-laden nature of all, 83
Ockham, William of, 50–51, 59
Organic philosophy, 44–45, 49–50,
    55, 69

Padmasambhava, and Great
    Perfection tradition, 109–112,
    115–118, 185
Panpsychism, 30
Paracelsus, and "inner illumination,"
    49–50. See also Organic
    philosophy
Perception(s). See also Mental
    perception
    versus assumption, expectation,
        and belief, 60–62
    as invariant on some fundamental
        level, 67
    paranormal abilities and
        extrasensory, 104–105, 108
    primary and secondary properties
        associated with, 34–35
    of relationships as well as discrete
        entities, 95–96
    theory-laden nature of, 95–96
    as type of experiential awareness
        with respect to objects of
        cognition, 92
Physicalism, 12, 25–27
    in early Greek thought, 42
    and marginalization of mind, 27–
        30
    roots of, in Christian theology, 51
Physics. See also Quantum theory
    manufactured nature of most
        phenomena of modern, 181

modern philosophical views on,
    20
Placebo effect, 167–168
Plato, 42, 185
Positivism. See also Scientific
    positivism
    and decline of introspection, 78
Protestantism. See also Christianity;
        Christian theology; Scientific
        materialism
    and critique of magic in Roman
        Church, 45
    and relation between Reformation
        and Scientific Revolution, 46–
        47, 55–56
    and theological sanction of
        scientific study of nature, 51–
        52
Psychology. See also Cognitive
        psychology; Cognitive science(s);
        Introspection; James, William;
        Mind; Wundt, Wilhelm
    German influence on American,
        77
    late nineteenth-century emergence
        of, 124
Psychometer, and purely
        technological detection of
        consciousness, 90, 130, 131
Pure consciousness. See under
        Consciousness
Pure experience. See under Radical
        Empiricism
Puritanism, and early scientific
        figures, 49
Putnam, Hilary, 63–67, 70, 118,
        185
    and conceptual schemes, 64–65
    and conventional-factual
        continuum, 63–65, 67
    and dualisms as functional rather
        than ontological, 64
    on relativity of existence and
        knowledge, 66
    and subjective-objective
        continuum, 64, 67, 118
    and theorizing-observing
        continuum, 66

Qualia. *See also* Consciousness; Mind;
  Scientific materialism;
  Subjectivity
  causal agency of, 141–143
  denial of existence of, 138–140
  ignoring of, in scientific
    materialism and cognitive
    psychology, 146–149
  reduction of, to physical processes,
    126–127, 158
  unanswered questions regarding,
    127
Quantum theory
  and causality, 69, 70–71, 82, 142–
    143
  as challenge to scientific
    materialism, 69–73, 80, 166–167
  consciousness, conceptual
    designation, and problem of
    measurement in, 69–72
  lack of impact of, on science as a
    whole, 143
  and nonlocality, 143
  symmetry in, 70–71

Radical Empiricism, 60, 62–64, 66–
  67, 112–115
  as holistic view of mental and
    physical as dependently related
    events, 75
  and hypothesis of direct
    acquaintance with reality, 63,
    65
  and mental/physical dualism as
    conceptual construct, 63
  perception as terminus of thought
    in, 60
  and principles of scientific
    materialism as hypotheses, 60
  and pure experience, 63, 112–115
  role of human experience in, 60,
    62–63
  truth in, 66–67
Rationality, as invariant on some
  fundamental level, 67
Reason, in sixteenth-century
  experimental philosophy, 44
Reductionism, 12, 23–24

doubts regarding, 24
as method and as belief, 23
rejection of principle of, in James's
  Radical Empiricism, 60
Reification
  of "field of consciousness" in
    James, 114
  of fundamental laws of physics, 20–
    21
  in Great Perfection tradition, 112
  of mind and matter, 81–82
  of subjective consciousness versus
    objective reality, 156
  of truths, 67
Religion, 7, 31–32, 185–186. *See also*
  Christianity; Christian theology;
  Durkheim, Emile;
  Protestantism
  benefits of scientific influence on
    comparative study of, 181
  contemporary American belief in,
    6, 27
  and interfaith dialogue, 11, 171–
    172
  and science, 3–8, 11, 36, 55, 164,
    185–188
  and scientific and religious
    discourse, 171–172
  and values and ideals, 7, 164
Religious studies, as academic
  department in state
  universities, 170
Renouvier, Charles, influence on
  William James of, 159

Scholastic realism, influence of, on
  modern science, 79–80
Schools, American, marginalization
  of religion in, 169–170
Science, 7–8, 17–19, 20. *See also*
  Inquiry; Scientific materialism;
  Scientific realism; Scientism;
  Theory
  conflation of scientific materialism
    and, 21, 37, 62, 145–158
  and need for new methodologies
    for study of consciousness, 75,
    88, 132–133

relevance of contemplative
traditions for, 11, 120, 178–184,
187–188
and religion, 3–8, 11, 36, 55, 164,
185–188
Scientific materialism, 12, 21–30. *See
also* Science
beginnings of, 32–33, 41–56
challenges to, 64, 69–73, 80, 142–
143, 166–168
conflation of science and, 21, 37,
62, 145–158
and exclusion of subjective
elements of human mind, 12–
13, 28–29, 35, 65, 88, 124–128,
133, 138, 142
and human happiness, 159–164
as impediment to scientific
advancement, 12–13, 56, 58–59,
88, 150–151, 158, 165, 166, 168,
184–187
institutionalization of, 164–175
as a religion, 13, 30–37, 169–171
religious influences on, 25–26, 33,
34, 49–52, 54
and sanitization of unscientific
aspects in lives of early
scientific figures, 56
worldview implicit in, 160–161,
164
Scientific method, 36
Scientific observation, 61–62
Scientific positivism, 37
Scientific realism, 17, 19–21
theological grounds of, 41
Scientific research, ideological
constraints on, 4, 129. *See also*
Scientific materialism
Scientific Revolution. *See* Mechanical
philosophy; Scientific
materialism
Scientific tradition, four dimensions
of, 13, 17. *See also individual
dimensions*
Scientism, 17, 37–39, 63
Scientists
belief in personal God among
American, 27

distinctions between scientific
materialists and, 32–33
as priests in the temple of nature,
51–52
Scot, Reginald, on witchcraft, 45
Searle, John
on fear of religion and "terror" of
subjectivity, 174
philosophy of mind of, 150–158
Self-monitoring, 168
Skepticism, in science, 17–18, 61–62
Skinner, B. F., on nonexistence of
mind and ideas, 28
Sorcery. *See under* Christianity, magic
and miracles
Soul, human, 4, 47–48, 53–54, 129–
130
Spiritual entities, 45, 52, 53, 54, 141–
142
Sprat, Thomas, in defense of
mechanical/experimental
philosophy, 52–54, 55, 142
Squires, Evan, on study of
consciousness, 120
Subjectivity, 5–6, 7, 35, 36, 47, 174.
See also Consciousness; James,
William; Mind; Objectivism;
Putnam, Hilary; Scientific
materialism; Taboo of
subjectivity
author's usage of the term, 21
scientific exclusion of, 12–13, 123–
127, 163
and subjectivity-objectivity
continuum, 67
Symmetry
central role of principles of, in
contemporary physics, 142–
143
historical progression from
assumptions of asymmetry to,
72

Taboo of subjectivity, 140, 174
as common to both religious
and scientific fundamentalism,
38
and introspection, 75, 87

Taboo of subjectivity (*continued*)
  and neglect of research into
      placebo effect, 168
  in religious studies departments of
      state universities, 170
Taboos, Durkheim and religious, 33–
      34
Tacit knowledge, in science and
      contemplative practice, 183
Technology, as means of survival and
      well-being, 7, 161, 163
Temperament, and choice of
      constructs and interpretations,
      20–21
Theology. *See also* Christian theology
  accommodation of scientific theory
      to current, 6
Theory
  "bootstrap" theory of validation of
      scientific, 66–67
  degenerate nature of asymmetric,
      72
  as describing only appearances
      and not reality, 20
  and inadequacy of theories of
      consciousness in Great
      Perfection tradition, 110
  influence of, on what is observed,
      83
  possibility of empirical refutation
      in scientific but not necessarily
      religious, 186
  testing of, in science, 18
Theory of imposed natural law, 50–
      51
Tibet, 9–10, 166
Tibetan Buddhism. *See* Buddhism,
      Tibetan
Trait-effects, lasting, as criteria for
      successful contemplative
      practice, 183
Truth(s)
  determination of, in science, 18
  invariant across all conceptual
      frameworks, 181
  in James's Radical Empiricism and
      in scientific practice, 66–67

of nature as inexorable and
      immutable, 35

Underdetermination, problem of, 20–
      21, 68–69, 94–95
Universalism, 12, 23, 46, 51

Values, and beliefs, 7, 164, 185
Vasubandhu, and Mahāyāna
      Buddhist contemplative
      tradition, 105–108
Voluntarist theories of natural and
      moral law, 51, 52–53

Wallace, B. Alan, scientific and
      religious background of, 8–
      11
Watson, John B., on discarding all
      subjective terms, 28
Weinberg, Steven, 142–143, 164
Weyer, Johann, on witchcraft, 45
Wheeler, John, on "observer-
      participancy" and human
      experience, 70
Whitehead, Alfred North, on
      relation between religion and
      science, 3
Wilson, Edward O., on religion and
      science, 25, 31–32
Witchcraft. *See under* Christianity,
      magic and miracles
World
  disenchantment of, 47, 56
  as independent of human mind,
      41
Worldview(s)
  conflict between mechanical and
      organic, 44–45
  implicit in principles of scientific
      materialism, 160–161, 164
  plurality of scientific, 19
  relation of values and ideals to
      Christian and materialist, 7,
      164
Wundt, Wilhelm, and early
      introspectionist school, 24, 77,
      79

CPSIA information can be obtained at www.ICGtesting.com
Printed in the USA
236062LV00002B/1/P

9 780195 173109